Lecture Notes in Computer Science   9260

Commenced Publication in 1973
Founding and Former Series Editors:
Gerhard Goos, Juris Hartmanis, and Jan van Leeuwen

Editorial Board

More information about this series at http://www.springer.com/series/8637

Abdelkader Hameurlain · Josef Küng
Roland Wagner · Alfredo Cuzzocrea
Umeshwar Dayal (Eds.)

Transactions on Large-Scale Data- and Knowledge- Centered Systems XXI

Selected Papers from DaWaK 2012

 Springer

Editors-in-Chief

Abdelkader Hameurlain
IRIT, Paul Sabatier University
Toulouse
France

Roland Wagner
FAW, University of Linz
Linz
Austria

Josef Küng
FAW, University of Linz
Linz
Austria

Guest Editors

Alfredo Cuzzocrea
ICAR-CNR and University of Calabria
Rende
Italy

Umeshwar Dayal
Hewlett-Packard Labatories
Palo Alto, CA
USA

ISSN 0302-9743   ISSN 1611-3349   (electronic)
Lecture Notes in Computer Science
ISBN 978-3-662-47803-5   ISBN 978-3-662-47804-2   (eBook)
DOI 10.1007/978-3-662-47804-2

Library of Congress Control Number: 2015943846

Springer Heidelberg New York Dordrecht London

Printed on acid-free paper

Springer-Verlag GmbH Berlin Heidelberg is part of Springer Science+Business Media
(www.springer.com)

Preface

Today, a great deal of attention is devoted to the issue of managing and mining big data, whose main goal consists in efficiently representing and extracting useful knowledge from such kind of data that encompass the well-known 3V characteristics, i.e., volume, velocity, and variety. This has occurred after it was recognized that traditional approaches developed during several years of data management and mining research are not suitable to comply with such novel characteristics. Another relevant property of big data to be considered is represented by their strict coupling with emerging cloud computing environments, which try to deal with research challenges deriving from managing and mining big data via specialized architectures, platforms, and paradigms based on the principles of high performance, high availability, and resource virtualization.

Within the broad scope of big data management and mining, data warehousing and knowledge discovery from big data plays a leading role and collects a wide family of models and methodologies for devising advanced data models (e.g., multidimensional models), warehousing, OLAPing, and extracting useful knowledge from big data, via a wide spectrum of specialized warehousing/mining "predicates," such as ETL processing, aggregation, data mart indexing, frequent pattern mining, machine learning techniques, emerging pattern mining, association rule discovery, etc. All these initiatives have a common denominator, i.e., starting from the limitations of traditional data warehousing and knowledge discovery approaches in dealing with big data, not being scalability issues is the only drawback to face-off.

Last but not least, data warehousing and knowledge discovery from big data also animates a very wide family of modern applications that, without doubt, are inspiring a plethora of novel models, techniques, and algorithms in this scientific context. Among others, relevant applications are: Web advertisement, scientific computing, social network data management, energy management systems, smart city applications, etc.

In order to fulfill the innovative requirements posed by the issue of realizing data warehousing and knowledge discovery in the big data era, this special issue on *Data Warehousing and Knowledge Discovery from Big Data* of *LNCS Transactions on Large-Scale Data- and Knowledge-Centered Systems* collects a selection of the best papers presented at 14th International Conference on Data Warehousing and Knowledge Discovery (DaWaK 2012), held in Vienna, Austria, during September 3–6, 2012. Following its successful tradition, DaWaK 2012 attracted a large number of submissions, and, after a rigorous selection process among the accepted conference papers, only 10 papers were invited for submission to the *LNCS Transactions on Large-Scale Data- and Knowledge-Centered Systems* special issue on *Advances in Data Warehousing and Knowledge Discovery*. After two thorough review rounds, only eight papers were accepted for final publication in this special issue.

The aim of the special issue is to offer an innovative, modern research perspective on the topic of data warehousing and knowledge discovery from big data, with particular emphasis on models, methods, and algorithms, by highlighting recent top-quality contributions and results in this scientific context, and, at the same, stimulating further investigation in the research field. In the following, we provide a summary of the papers included in the special issue.

The first paper, titled "Efficient Level-Based Top–Down Data Cube Computation Using MapReduce," by Suan Lee, Jinho Kim, Yang-Sae Moon, and Wookey Lee, identifies data cubes as essential parts of OLAP to support efficient multi-dimensional analysis over large-size data. The computation of data cube takes relevant amounts of time, because a data cube with d dimensions consists of 2d (i.e., exponential order of d) cuboids. To build ROLAP (Relational OLAP) data cubes efficiently, many algorithms (e.g., GBLP, Pipe-Sort, Pipe-Hash, BUC, etc.) have been developed, which share sort cost and input data scans in order to reduce data computation time. Several parallel processing algorithms have also been proposed. On the other hand, MapReduce is recently emerging as an authoritative framework for processing huge volumes of data, such as Web-scale data, in a distributed/parallel manner via using a large number of computers (e.g., several hundred or thousands). In the MapReduce framework, the degree of parallel processing is more important than elaborate strategies (e.g., short-share and computation-reduction) in order to reduce total execution time. Following these main considerations, the authors propose two distributed parallel processing algorithms. The first one, called MRLevel, heavily borrows from the inherent features of the MapReduce framework. The second one, called MRPipeLevel, is based on the existing Pipe-Sort algorithm that is one of the most efficient for supporting top–down cube computation. The MRLevel algorithm tries to parallelize cube computation and to reduce the number of data scans by level at the same time. The MRPipeLevel algorithm is based on the functionalities and benefits of MRLevel, and aims at further reducing the number of data scans by pipelining at the same time. Finally, the authors also identify factors for enhancing the performance of MapReduce in order to process very huge data.

The second paper, titled "Differentiated Multiple Aggregations in Multi-dimensional Databases," by Ali Hassan, Frank Ravat, Olivier Teste, Ronan Tournier, and Gilles Zurfluh, focuses on multidimensional databases (MDBs), which support efficient querying and analysis of data stored in a data warehouse. In this context, classical MDBs support only the calculation of a measure made by the same aggregation function while performing drilling or rotating operations (i.e., changing the analyzed slice of the underlying data cube). For instance, if we consider sales amounts, these can be calculated as the sum of the products sold by cities and years. When drilling from cities to countries, the new amounts are calculated using the same aggregation function. When the user wishes to change the aggregation function between two slices of the manipulated cube, classical MDBs no longer guarantee the validity of the calculated data, or even worse: They do not support this type of manipulation. In order to provide solutions to this limitation, the authors provide a novel conceptual model that supports (1) multiple aggregations, which associate to the same measure a different aggregation function according to analysis hierarchies, and (2) differentiated aggregation, which allows for specific aggregations at each detail level. The proposed model is based on a graphical formalism that allows one to control the validity of aggregation functions (distributive, algebraic, or holistic). Finally, the authors also show how conceptual modeling can be used in a ROLAP environment in order to build lattices of pre-computed aggregates.

The third paper, titled "MIRABEL DW: Managing Complex Energy Data in a Smart Grid," by Laurynas Šikšnys, Christian Thomsen, and Torben Bach Pedersen, presents research and practical results from the MIRABEL project, which focuses on the definition and development of a data management system for smart grids targeted at achieving smarter scheduling of energy consumption such that, for instance, charging of car batteries is done during the night when there is an overcapacity of green energy from windmills etc. Energy can then be requested by means of flex-offers which define flexibility with respect to time, amount, and/or price. The authors describe MIRA-BEL DW, a data warehouse (DW) for the management of the large amounts of complex energy data in the MIRABEL system. In more detail, they present a unified schema that can manage data both at the level of the entire electricity network and the level of individual nodes, such as a single consumer node. The schema has a number of complexities compared with typical DW schemas. These include facts about facts and composed non-atomic facts and unified handling of different kinds of flex-offers and time series. The authors also discuss alternative data modeling strategies and how specialized variants of the generic schema can be used by different node types while maintaining compatibility and consistency between them. Finally, the authors complement their analytical contributions by presenting typical queries from the energy domain, and a related performance study.

The fourth paper, titled "Modular Neural Networks for Extending OLAP to Prediction," by Wiem Abdelbaki, Sadok Ben Yahia, and Riadh Ben Messaoud, takes into consideration limitations of classical OLAP analysis that, as the authors recognize, offers a good applications package to explore and navigate data cubes, but, unfortunately, it is limited to exploratory tasks. As a consequence, OLAP does not assist the decision maker in performing information investigation. Thus, various studies have been trying to extend OLAP to new capabilities by coupling it with data-mining algorithms. The paper stands within this trend, and introduces two major contributions. First, a multi-perspectives cube exploration framework (MCEF) is introduced. MCEF is a generalized framework designed to assist the application of classical data-mining algorithms on OLAP cubes. Second, a neural approach for prediction over high-dimensional cubes (NAP-HC) is also introduced, which extends modular neural networks (MNN) architecture to the multidimensional context of OLAP cubes, to predict non-existent measures. A pre-processing stage is embedded in NAP-HC to assist it in facing the challenges arising from the particularity of OLAP cubes. This phase consists of an OLAP-oriented cube exploration strategy coupled with a dimensionality reduction step that replies on principal component analysis (PCA). The experiments described highlight the efficiency of MCEF in assisting the application of MNN on OLAP cubes and the high predictive capabilities of NAP-HC.

The fifth paper, titled "Cut-and-Rewind: Extending Query Engine for Continuous Stream Analytics," by Qiming Chen and Meichun Hsu, focuses on combining data warehousing and stream processing technologies, which has proved to have great potential in offering low-latency data-intensive analytics. Unfortunately, such convergence has not been properly addressed so far. The current generation of stream-processing systems is in general built separately from the data warehouse and query engine, which can cause significant overhead in data access and data movement, and is unable to take advantage of the functionalities already offered by the existing data warehouse systems. Starting from this evidence, the authors tackle some hard problems in integrating stream analytics capability into the existing query engine. They introduce an extended SQL query model that unifies queries over both static relations and dynamic streaming data, and they develop techniques to extend query engines to support the unified model. Also, they propose the cut-and-rewind query execution model to allow a query with full SQL expressive power to be applied to stream data by converting the latter into a sequence of "chunks," and executing the query over each chunk sequentially, but without shutting the query instance down between chunks for continuously maintaining the application context across the execution cycles as required by sliding-window operators. They also propose the cycle-based transaction model to support continuous querying with continuous persisting (CQCP) with cycle-based isolation and visibility. In order to support their framework, the authors finalize the implementation of their approach by extending the PostgreSQL, thus resulting in a new kind of tightly integrated, highly efficient system with advanced stream-processing capability as well as full DBMS functionality. The authors demonstrate the system with the popular linear road benchmark, and report on the performance. By leveraging the matured code base of a query engine to the maximal extent, the proposed approach can significantly reduce the engineering investment needed for developing the streaming technology.

The sixth paper, titled "Mining Popular Patterns: A Novel Mining Problem and Its Application to Static Transactional Databases and Dynamic Data Streams," by Alfredo Cuzzocrea, Fan Jiang, Carson K. Leung, Dacheng Liu, Aaron Peddle and Syed K. Tanbeer, recognizes that, since the introduction of the frequent pattern mining problem, researchers have extended frequent patterns to different useful patterns such as cyclic, emerging, periodic, and regular patterns. In line with this trend, the paper introduce popular patterns, which captures the popularity of individuals, items, or events among their peers or groups. Moreover, they also propose the Pop-tree structure for capturing the essential information from transactional databases, and the Pop-growth algorithm for mining popular patterns from the Pop-tree. The authors illustrate how the proposed algorithm mines popular friends from social networks, as a relevant case study of the proposed framework. Because the framework is not confined to mining popular patterns from static transactional databases, they extend the work to mining popular patterns from dynamic data streams. Specifically, the Pop-stream structure to capture the popular patterns in batches of data streams is proposed, as well as the Pop-streaming algorithm for mining popular patterns from the Pop-stream structure. Finally, the experimental results show that (a) the proposed tree structure is compact and space efficient and (b) the proposed algorithm is time efficient in mining popular patterns from static transactional databases and dynamic data streams.

The seventh paper, titled "Rare Pattern Mining from Data Streams Using SRP-Tree and Its Variants," by David Tse Jung Huang, Yun Sing Koh, and Gillian Dobbie, addresses research in the area of rare pattern mining where the researchers try to capture patterns involving events that are unusual in a data set. These patterns are considered more useful than frequent patterns in some domains, including detection of computer attacks or fraudulent credit transactions. To date, most of the research in this area has concentrated only on finding rare rules in a static data set. Nevertheless, there is a proliferation of applications that generate data streams, such as network logs and banking transactions, and applying techniques that mine static data sets is not practical for data streams. In order to fill this gap, the authors propose a novel approach called streaming rare pattern tree (SRP-Tree) and its variations, which finds rare rules in a data stream environment using a sliding window, and show that it both finds the complete set of item sets and runs with fast execution time.

Finally, the eight paper, titled "Improving Cross-Document Knowledge Discovery Through Content and Link Analysis of Wikipedia Knowledge," by Peng Yan and Wei Jin, focuses on the research context of the vector space model (VSM), which has been widely used in natural language processing (NLP) for representing text documents as a bag of words (BOW). However, according to this model, only document-level statistical information is recorded (e.g., document frequency, inverse document frequency) and word semantics cannot be captured. Improvement in understanding the meaning of words in texts is a challenging task and sufficient background knowledge may need to be incorporated to provide a better semantic representation of texts. Following this main trend, the authors present a text-mining model that can automatically discover semantic relationships between concepts across multiple documents, where the traditional search paradigm such as search engines cannot help much, and effectively integrate various evidence mined from Wikipedia knowledge. The authors argue that this integration may effectively complement existing information contained in text corpus and facilitate the construction of a more comprehensive representation and retrieval framework. Experimental results demonstrate that the search performance has been significantly enhanced when compared with two competitive baseline methods.

The editors would like to express their sincere gratitude to the Editors-In-Chief of LNCS Transactions on Large-Scale Data- and Knowledge-Centered Systems, Prof. Abdelkader Hameurlain, Prof. Josef Küng, and Prof. Roland Wagner, for accepting their proposal of a special issue focused on data warehousing and knowledge discovery from big data, and for assisting them whenever required. The editors would also like to thank all the reviewers who have worked within a tight schedule and whose detailed and constructive feedbacks to authors have contributed to substantial improvement in the quality of the final papers.

June 2015   Alfredo Cuzzocrea
                                                                             Umeshwar Dayal

Organization

Editorial Board

Additional Reviewers

Filippo Furfaro   University of Calabria, Italy
Pedro Furtado   University of Coimbra, Portugal
Jens Lechtenboerger   Munster University, Germany
Carson Leung   University of Manitoba, Canada
Elio Masciari   ICAR-CNR, Italy
Antonio Rinaldi   University "Federico II" of Naples, Italy
Robert Wrembel   Poznan University of Technology, Poland

Contents

Efficient Level-Based Top-Down Data Cube Computation
Using MapReduce... 1
    Suan Lee, Jinho Kim, Yang-Sae Moon, and Wookey Lee

Differentiated Multiple Aggregations in Multidimensional Databases....... 20
    Ali Hassan, Frank Ravat, Olivier Teste, Ronan Tournier,
    and Gilles Zurfluh

MIRABEL DW: Managing Complex Energy Data in a Smart Grid........ 48
    Laurynas Šikšnys, Christian Thomsen, and Torben Bach Pedersen

Modular Neural Networks for Extending OLAP to Prediction............ 73
    Wiem Abdelbaki, Sadok Ben Yahia, and Riadh Ben Messaoud

Cut-and-Rewind: Extending Query Engine for Continuous Stream
Analytics... 94
    Qiming Chen and Meichun Hsu

Mining Popular Patterns: A Novel Mining Problem and Its Application
to Static Transactional Databases and Dynamic Data Streams............ 115
    Alfredo Cuzzocrea, Fan Jiang, Carson K. Leung, Dacheng Liu,
    Aaron Peddle, and Syed K. Tanbeer

Rare Pattern Mining from Data Streams Using SRP-Tree and Its Variants ... 140
    David Tse Jung Huang, Yun Sing Koh, and Gillian Dobbie

Improving Cross-Document Knowledge Discovery Through Content
and Link Analysis of Wikipedia Knowledge........................ 161
    Peng Yan and Wei Jin

Author Index 185

Efficient Level-Based Top-Down Data Cube Computation Using MapReduce

Suan Lee[1], Jinho Kim[1(✉)], Yang-Sae Moon[1], and Wookey Lee[2]

[1] Department of Computer Science, Kangwon National University,
192-1 Hyoja-dong, Chuncheon, Kangwon, Korea
{salee, jhkim, ysmoon}@kangwon.ac.kr
[2] Department of Industrial Engineering, Inha University,
100 Inha-ro, Nam-ku, Incheon, Korea
wookeylee@gmail.com

Abstract. Data cube is an essential part of OLAP(On-Line Analytical Processing) to support efficiently multidimensional analysis for a large size of data. The computation of data cube takes much time, because a data cube with d dimensions consists of 2^d (i.e., exponential order of d) cuboids. To build ROLAP (Relational OLAP) data cubes efficiently, many algorithms (e.g., GBLP, PipeSort, PipeHash, BUC, etc.) have been developed, which share sort cost and input data scan and/or reduce data computation time. Several parallel processing algorithms have been also proposed. On the other hand, MapReduce is recently emerging for the framework processing huge volume of data like web-scale data in a distributed/parallel manner by using a large number of computers (e.g., several hundred or thousands). In the MapReduce framework, the degree of parallel processing is more important to reduce total execution time than elaborate strategies like short-share and computation-reduction which existing ROLAP algorithms use. In this paper, we propose two distributed parallel processing algorithms. The first algorithm called MRLevel, which takes advantages of the MapReduce framework. The second algorithm called MRPipeLevel, which is based on the existing PipeSort algorithm which is one of the most efficient ones for top-down cube computation. (Top-down approach is more effective to handle big data, compared to others such as bottom-up and special data structures which are dependent on main-memory size.) The proposed MRLevel algorithm tries to parallelize cube computation and to reduce the number of data scan by level at the same time. The MRPipeLevel algorithm is based on the advantages of the MRLevel and to reduce the number of data scan by pipelining at the same time. We implemented and evaluated the performance of this algorithm under the MapReduce framework. Through the experiments, we also identify the factors for performance enhancement in MapReduce to process very huge data.

Keywords: Data cube · ROLAP · MapReduce · Hadoop · Distributed parallel computing

© Springer-Verlag Berlin Heidelberg 2015
A. Hameurlain et al. (Eds.): TLDKS XXI, LNCS 9260, pp. 1–19, 2015.
DOI: 10.1007/978-3-662-47804-2_1

1 Introduction

Due to the advance of information technology and WWW(World-Wide Web) recently, many applications require to manage a large amount data and to analyze them on-line over multi-dimensions. In order to handle these requirements efficiently, there have been a lot of researches on multidimensional data cubes [1]. The data cube is an essential part of OLAP(On-Line Analytical Processing), which maintains aggregate results pre-computed over source data sets. It takes a lot of time to compute a data cube., it takes a lot of time because each data cube keeps the values aggregated by every possible combination of dimensional attributes. If a source table have T tuples, the cost of $T \times 2^D$ is required to compute a data cube with D dimensions. In order to reduce such high cost problem of the multidimensional cube computation, many algorithms have been proposed [2–5]. These algorithms are classified into several categories such as Relational OLAP(ROLAP), Multidimensional OLAP (MOLAP), and Graph-Based methods [1].

This paper focuses on ROLAP cube computation because it can be easily incorpo-rated into existing DBMSs. GBLP [1], PipeSort [3], PipeHash [3], and BUC [5] are examples of ROLAP cube computation. These algorithms reduce cube computation time by sharing sort cost and input data scan and/or by reducing data computation. Some others proposed parallel processing algorithms [6–9] e.g., RP, BPP, ASL [9], and PnP [7]. While these algorithms use paralleling processing computer/cluster consisting of several ten CPUs, MapReduce which are recently emerging can use a large number of computers (e.g., several hundred or thousands more). It becomes a popular framework to handle efficiently huge volume of data like web-scale data in distributed parallel manner. To reduce cube computation time, several algorithms [14–17] (e.g., MR-Cube [17]) based on the MapReduce framework have been also developed. However these algorithms use bottom-up approach to compute closed cubes and/or data cubes on holistic measurements. Compared to the top-down approach, the bottom-up approach is difficult to handle very huge data set and it is limited to utilize parallel processing, because it should load a set of data into main memory.

Recently, it is required to analyze and to manage extremely large data such as Web data and social media. In order to handle these massive data, this paper proposes MRLevel and MRPipeLevel, which are distributed parallel data cube computation algorithm to efficiently build massive data cube with the MapReduce framework. The MRLevel calculates cuboids which are in the same level of a cube lattice and should be sorted. The MRPipeLevel is based on the existing PipeSort algorithm which is known as one of the most efficient ones for top-down ROLAP cube computation. This method reduces the number of data scan by pipelining the computation of several cuboids with the same sorting order at the same time.

In this paper, we implement and evaluate the proposed algorithms through various experiments. We carry out a diversity of experiments with large scale high-dimensional data, and comparative experiments with the MRNaïve, MRGBLP and MRPipeSort (i.e., the MapReduce version of Naïve, GBLP and PipeSort algorithms respectively) which are typical top-down ROLAP data cube algorithms. Through the experiments, we found that the proposed algorithm is more efficient than others and effective to handle large-scale multidimensional data through the MapReduce framework. We also identify the important factors for performance enhancement in MapReduce to process very huge data.

2 Background

2.1 Data Cube

A data cube consists of measurements and dimensions which are the data to analyze and the criteria for analysis, respectively. The cube keeps aggregate values for the GROUP BYs of every possible combination of dimensions. The result for each GROUP BY is called a cuboid, and all of cuboids forms a lattice structure according to their inclusion relationship. Figure 1 shows a cube lattice structure built for sales data by year, store and item dimensions.

Aggregate functions which calculate aggregate values stored in the cells of a data cube, can be classified into three types as follows:

- Distributive: COUNT(), MIN(), MAX(), SUM()
- Algebraic: AVG(), standard deviation, MaxN(), MinN(), center_of_mass()
- Holistic: Median(), Mode(), Rank()

Among them, distributive and algebraic functions could compute lower cuboids by using upper cuboids of a cube lattice structure. In Fig. 1, the <Year> cuboid can be computed from the <Year, Store> cuboid. For example, if they use measurements as SUM(Sales), the <Year, Store> cuboid has <2012, S1, *, 100> , <2012, S2, *, 81> , <2011, S1, *, 18> , <2011, S2, *, 57> , and <2011, S3, *, 32> cells. These cells can be used to find out <2012, *, *, 181> and <2011, *, *, 107> cells for the <Year> cuboid. With this inclusion relationship between cuboids, a cuboid can be computed from several cuboids in its upper level of a cube lattice structure. By taking advantage of this inclusion relationship, the cube computation time can be reduced. Several top-down have been developed by using this concept.

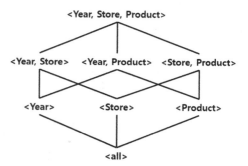

Fig. 1. Examples of source data and cube lattice

2.2   MapReduce

MapReduce is a distributed parallel processing technology for massive data, which was proposed by Google in 2004 and has been applied to various services of Google. The MapReduce are being used in various applications and it becomes a standard practically in large-scale parallel processing fields. To handle massive datasets in the MapReduce framework, in addition, a distributed file system is needed. GFS (Google File System) [10] and HDFS(Hadoop Distributed File System) [12] are popular distributed file systems. The GFS was developed by Google and the HDFS by an open source software development project, the Hadoop [11], which uses similar architecture and functions to GFS. This paper utilizes Hadoop's MapReduce and HDFS.

As shown in Fig. 2, the data flow of the MapReduce is as follows: (1) input data is split to deliver to map functions; (2) each map function stores the split input data into its own in-memory buffer, and it partitions, sorts, and spills the input data into disks; (3) the copy phase merges the partitions in the result of each map function; (4) the sort phase delivers the merged results to corresponding reduction functions; and (5) each reduce function processes the delivered data and output its final result to HDFS.

3   The MRLevel Algorithm

3.1   Cube Execution Tree

The MRLevel is an algorithm proceeds to the level unit, a top-down cube algorithm. For example, the MRLevel builds a cube execution tree of Fig. 3(a) from a cube lattice structure of Fig. 3(a) to processes level by level. The execution tree consists of the cuboids which is smallest parents cuboids with theirs child cuboids. To create a cube execution tree, we need sizes of cuboids but it takes additional cost to obtain the size of data and cardinality. Therefore, the MRLevel does not perform a pre-calculated to calculate the size of cuboids and builds a whole cube execution tree. The MRLevel stores cuboid result with cuboid size when performing calculations in each cuboid by level from cube lattice structure. The MRLevel can be selected the smallest parent using cuboid size when calculating theirs child cuboids.

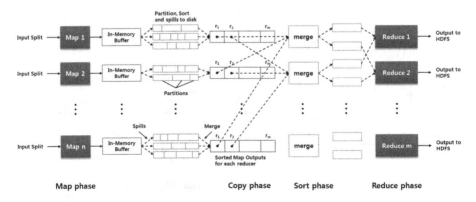

Fig. 2. The data flow of MapReduce

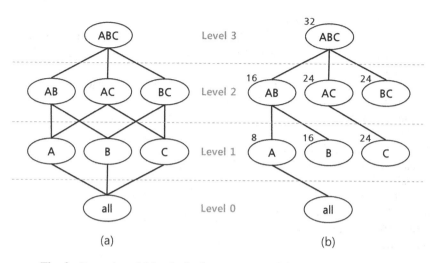

Fig. 3. Examples of (a) cube lattice structure and (b) cube execution tree

For example, Fig. 3 includes parent cuboids and their child cuboids with cuboid size. A cuboid A is selects smallest cuboid AB as a parent between cuboid AB and AC. Cuboid B is also selects smallest cuboid AB between cuboid AB and cuboid BC. In the case of cuboid C, It can be a parent cuboid with cuboid AB or cuboid BC. However, if these are the same as the cuboid size, it may choose one anything which of two cuboids. Thus, the MRLevel algorithm calculates child cuboid by each level using selected smallest parent cuboid.

3.2   Shared Scanning and Reduction of MapReduce Phases

The MRLevel algorithm shares scan cost and performs MapReduce phase as much as number of levels. A cuboid ABC in level 1 as input, cuboid AB, AC, and BC emits when it scan the input data once at Fig. 3. In other words, the MRLevel once read cuboid ABC, and then write cuboid AB, AC, and BC so its cost is reduced. In addition, the MRLevel algorithm uses multiple input data in input part and emits multiple output data in output part. For example, cuboid A and B in level 2 used cuboid AB as input, and cuboid C used cuboid AC as input. The MRLevel algorithm emits cuboid A, B, and C in level 1 using cuboid AB and AC as input in level 2.

3.3   Computation Cost of MRLevel

The cost of the MRLevel algorithm equals to the sum of the cost in each level. In Fig. 3, for example, when cuboid ABC calculates cuboid AB, AC, and BC, we can be calculated as a cost: read(ABC) + write(AB) + read(ABC) + write(AC) + read (ABC) + write(BC). The MRLevel calculates cuboid AB, AC, and BC on the same level from cuboid ABC at the same time. It presents the cost of the calculation: read (ABC) + write(AB) + write(AC) + write(BC). The results of the estimated cost of basic data cube and the MRLevel using the size of each cuboid in Fig. 3 is as follows:

- The cost of basic data cube:
  = read(ABC) + write(AB) + read(ABC) + write(AC) + read(ABC) + write(BC)
  + read(AB) + write(A) + read(AB) + write(B) + read(AC) + write(C)
  + read(A) + write(all)
  = read(32) + write(16) + read(32) + write(24) + read(32) + write(24)
  + read(16) + write(8) + read(16) + write(16) + read(24) + write(24)
  + read(8) + write(1)
  = 273
- The cost of the MRLevel data cube:
  = read(ABC) + write(AB) + write(AC) + write(BC)
  + read(AB) + write(A) + write(B) + read(AC) + write(C)
  + read(A) + write(all)
  = read(32) + write(16) + write(24) + write(24)
  + read(16) + write(8) + write(16) + read(24) + write(24)
  + read(8) + write(1)
  = 193

The MRLevel algorithm reduces the cost about 30 % than the basic data cube where the cost of reading and writing are assumed to be the same cost.

3.4 MapReduce Data Flow of MRLevel

Figure 4 is example of data flow about the MRLevel algorithm processing. In the figure, the MRLevel executes a total of four MapReduce phases about three-dimensional input data. In the first MapReduce phase, the MRLevel calculates cuboid ABC using raw data as input. We use the key on the value of each cell, and use 1 value as the measure value for obtaining function COUNT(). The MRLevel emits in the form of the <cell, value>, and then merges cell with the same value. For example, in the map() function, the MRLevel emits data <1 1 1, 1> and <1 1 1, 1> and merges with <1 1 1, [1]> in shuffle phase. Then emits calculated data <1 1 1, 2> to reduce() function. It calculates a cuboid ABC as a result of the <1 1 1, 2>, <1 1 3, 1>, and <1 2 2, 1>.

In the second MapReduce phase, the MRLevel emits cells of cuboid AB, AC, and BC using a cuboid ABC as input in map() function and calculates each cuboid AB, AC, and BC. In the third phase, as a cube execution tree in Fig. 3, the MRLevel emits cells of cuboid A and B by selecting smallest parent cuboid AB and emits cells of cuboid C using cuboid AC. Then, the MRLevel calculates cells of each cuboid in reduce() function. Finally, in the fourth phase of MapReduce, the MRLevel calculates cuboid All. It can be calculated any parent cuboid. Thus, the MRLevel calculates cuboid All using smallest cuboid A as input data.

3.5 MRLevel Algorithm

The MRLevel algorithm separates by Map(), Reduce(), and MRLevel() function shown in Fig. 5. In the MRLevel() function, it call Map() function to emit top cuboid from raw data and call Reduce() function for calculating measure value. The MRLevel declares set E of execution cuboids and traversals to all the cuboid in level k + 1.

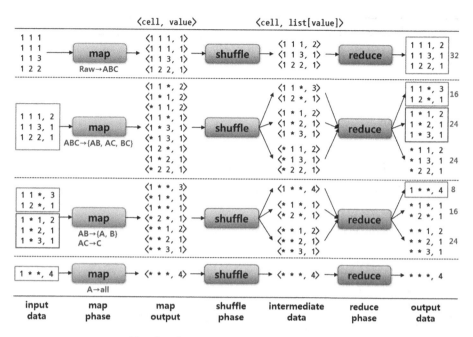

Fig. 4. The example of MRLevel data flow

The MRLevel progresses by step in level k of cube execution tree. The MRLevel finds candidates which can be a parent cuboid in level k + 1. The MRLevel declares parent cuboid M as a smallest cuboid of candidates in parent cuboid P. A parent cuboid M and child cuboid C are inserted into the set E. For example, the MRLevel stores the form of AB → {A, B}, AC → C in E when it browses a set of execution cuboid for level 2 in Fig. 3. A Map() function takes as input a set E of execution cuboid, and scans every child cuboid of parent cuboid in a set E. At this time, the MRLevel emits (cell, measure) pair of child cuboid C from (cell, measure) pair of parent cuboid M. For example, the MRLevel emits a cell <1 * *, 3> in child cuboid A from a cell <1 1 *, 3> in parent cuboid AB. A Reduce() function merges same cell of released data from Map() function and aggregates all measure according to a user-specified function such as COUNT(). For example, cell <1 * *, 3> and <1 * *, 1> are merged, then it comes in the form <1 * *, (3, 1)> in Reduce() function. If a user-specified function is COUNT(), a Reduce() function emits cell <1 * *, 4> using function COUNT(3, 1).

4   MRPipeLevel: An Integration of MRLevel and PipeSort

The MRPipeLevel is an algorithm based on the PipeSort, a top-down cube computation algorithm. The PipeSort generates the minimum cost sort plan tree from a cube lattice which represents the cuboids to be sorted in corresponding orders and the other cuboids sharing the sort result of each sorted cuboids to minimize the total cube computation time. It computes a set of cuboids sharing the same sort order together with one scan of

Algorithm   MRLevel
Map()
Input
E is execution cuboid set
Output
$\langle cell, measure \rangle$
Description
1 **for each** parent cuboid M in E
2   **for each** child cuboid C in M
3      $C(cell, measure) \leftarrow M(cell, measure)$ // $\langle 1 * *, 3 \rangle \leftarrow \langle 1\ 1\ *, 3 \rangle$
4   **emit** $\langle cell, measure \rangle$
Reduce()
Input
$\langle cell, measures\,(m_1, m_2, \cdots, m_n) \rangle$
Output
$\langle cell, measure \rangle$
Description
1 $measure = function(m_1, m_2, \cdots, m_n)$
2 **emit** $\langle cell, measure \rangle$
MRLevel()
Description
1 $R = R \cup \boldsymbol{Map}([\text{raw data, top cuboid}]) \cup \boldsymbol{Reduce}()$
2 **for each** level k in cube lattice
3   execution cuboid set $E = [\emptyset, \emptyset]$
4   **for each** cuboid C in k + 1
5      $P = C's$ candidate parent cuboids in k
6   minimum parent cuboids $M = \boldsymbol{minimum}(\{p_1, p_2, \cdots p_m\} \in P)$
7      $E = E \cup [M, C]$ // $AB \rightarrow \{A, B\}, AC \rightarrow C$
8    $R = R \cup \boldsymbol{Map}(E) \cup \boldsymbol{Reduce}()$
9 **return** R

Fig. 5. The algorithm of MRLevel

source table (or another cuboid) by pipelining the computation of these cuboids. The proposed MRPipeLevel incorporates a distributed parallel processing strategy for the PipeSort in the MapReduce framework which maximizes the degree of parallelism and minimizes the number of MapReduce phases and the number of data scans.

4.1   Sort Tree with Pipeline

The MRPipeLevel builds sort trees and pipelines from a cube lattice structure to compute data cubes efficiently. The sort trees represent the cuboids which don't share the sort order of their parent cuboids thus have to sort these parents to compute them. For example, Fig. 6 includes two sort trees whose root nodes are represented as dotted circles (i.e., ABC and AC cuboids with dotted circles). The tree in the middle of the

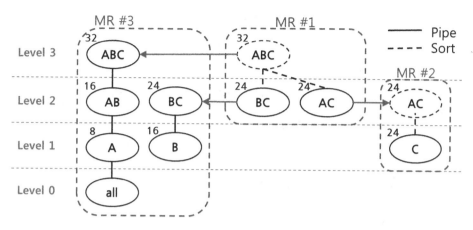

Fig. 6. The example of cube execution pipe tree

figure shows that the cuboid BC is computed by sorting the cuboid ABC in the order of AB and the cuboid AC by sorting the ABC in the order of AC. In order to reduce the number of MapReduce phases and to maximize the degree of parallelism, the MRPipeLevel processes each sort tree level by level which all the cuboids in the same level are sorted together from their parents by one MapReduce phase. Thus two MapReduce phases are used to process the sort trees in the example of Fig. 6. After computing the cuboids in sort trees, the MRPipeLevel executed the other cuboids with pipelines.

4.2   Pipelines

Pipeline Aggregation. The MRPipeLevel's pipeline technique is a method used in the PipeSort algorithm [3], which computes several child cuboids without sorting if they correspond to their parent cuboid's prefix. As shown in Fig. 7, the AB, A, and all cuboids can be calculated in the course of calculating the ABC cuboid by the pipeline technique, without sorting and scanning input data repeatedly. When carrying out pipeline aggregation on the first tuple, for example, it computes ABC: <1 1 1, 1>, AB: <1 1 *, 1>, A: <1 * *, 1>, all: <* * *, 1>, and for the second tuple, it does ABC: <1 1 1, 2>, AB: <1 1 *, 2>, A: <1 * *, 2>, all: <* * *, 2>. If aggregating the third

Input data

1	1	1
1	1	1
1	1	3
1	2	2

COUNT()

Pipeline: ABC→AB→A→all

ABC	AB	A	all
<1 1 1, 2>			
	<1 1 *, 3>		
<1 1 3, 1>		<1 * *, 4>	<* * *, 4>
<1 2 2, 1>	<1 2 *, 1>		

Fig. 7. The example of pipeline aggregation

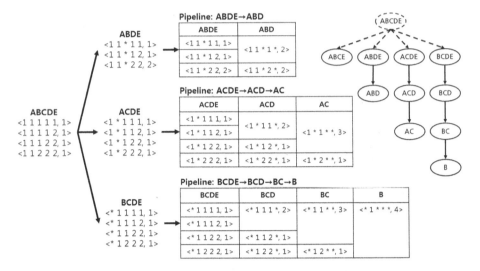

Fig. 8. The example of multi-pipeline aggregation

tuple, <1 1 1, 2> is emitted as a resulting cell of the ABC cuboid and it produces ABC: <1 1 3, 1>, AB: <1 1 *, 3>, A: <1 * *, 3>, all: <* * *, 3>. When accepting the fourth tuple, <1 1 3, 1> and <1 1 *, 3> are emitted as the resulting cells of ABC and AB cuboids respectively. If we use such a pipeline aggregation, all cells of the ABC, AB, A, and all cuboids can be computed together by scanning input data once. The MRPipeLevel minimizes the computation time by exploiting this pipeline aggregation by distributed processing within a MapReduce phase.

Multi-Pipeline Aggregation. In the MRPipeLevel, multiple pipelines can be simultaneously executed within one MapReduce phase. Figure 8 shows an example sort plan tree for a 5-dimensional cube. It contains three aggregation pipelines starting from ABCE, ABDE, ACDE, BCDE cuboids, while these four cuboids can be computed from the top cuboid ABCDE. The MRPipeLevel executes all of these three pipelines in parallel by one MapReduce phase. That is, ABDE, ABD, ACDE, ACD, AC, BCDE, BCD, BC and B cuboids are computed by one MapReduce phase to maximize the degree of parallelism.

4.3   Computation Cost of MRPipeLevel

The MRPipeLevel algorithm is similar to the MRLevel. The cost of the MRPipeLevel is as same as which excludes the cost of connected to the cuboid through pipe in the MRLevel. For example in Fig. 6, the cost of write(ABC) includes the cost of write (AB + A + all) because cuboid ABC is connected to the cuboid AB, A and All in the pipe. In other words, cuboid AB, A, and All are calculated when calculating cuboid ABC. The estimated cost using example are as follows:

- The computational cost of the cube of the MRPipeLevel
  = write(ABC) included (AB + A + all)
  + write(BC) included (B) + write(AC) included (C)
  = read(32) included (16 + 8 + 1)
  + write(24) included (16)
  + write(24) included (24)
  = 80

The cost of the MRPipeLevel reduced the cost about 40 % than the cost of MRLevel. We read and write costs are considered equal.

4.4 Handling Big Data and High Dimensional Cube

Recently, with an explosive increase of data, big data is required for multidimensional analysis. Because the algorithm proposed in this paper fundamentally exploits the MapReduce, it could flexibly cope with big data through scalability that increases the number of clusters according to the size of big data. Due to a problem such as costs of adding clusters, however, there may be a limit to the number of clusters. In addition, for data with a large dimension, the size of cubes is increased exponentially. The MRPipeLevel carries out the MapReduce phases level by level. In a cube lattice structure, the largest number of cuboids exists at an intermediate level of the lattice. For a data cube with very high dimension, the cuboids in a level cannot be computed together by one MapReduce phase because lots of data will be emitted at the same time.

This paper considers how to effectively control the emission of data that could not be computed at a MapReduce phase when computing cubes for big data and high-dimensional data as follows.

- If there are several sort sub-trees at each level, it divides the MapReduce phase for each sub-tree. Because the scanned cuboids are not the same, it could compute without additional costs.
- If there are lots of children cuboids emitted at the same time, it divides them to carry out the MapReduce phase. For such a case, the cost to scan a parent cuboid repetitively is added.
- If data is large enough to be difficult even to emit a cuboid, it partitions data itself to carry out the MapReduce phase. In such a case, there is no additional scan cost, however, the number of sort sub-trees and MapReduces is increased as much as the number of partitions and additional cost is incurred as it carries out the process to merge results of cuboids divided into each partition.

4.5 MapReduce Data Flow of MRPipeLevel

Figure 9 is an example for a data flow in the process carried out by the MRPipeLevel. Looking at the figure, a total of three MapReduce processes are carried out for three-dimensional input data. At the first MapReduce phase, the map function emits a cell corresponding to the ABC cuboid for the original data. The emitted cells of the ABC cuboid are aggregated for the same cell. At the second MapReduce phase, the map function uses the ABC cuboid as an input to emit cells of the AC and BC cuboids.

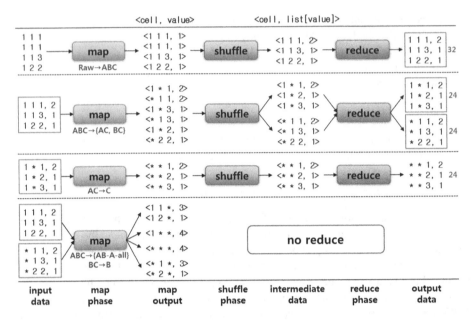

Fig. 9. The example of MRPipeLevel data flow

The reduce function computes the AC and BC cuboids to emit them. At the third MapReduce phase, the map function uses the AC cuboid as an input to emit the C cuboid's cell, and the reduce function computes the C cuboid to emit. At the fourth MapReduce phase, it computes the AB-A-all and B cuboids, which comprise of a pipeline, to emit. However, because the ABC and BC cuboids used as input data are the sorted data, the reduce function is not operated and the result could be computed immediately.

4.6   MRPipeLevel Algorithm

The MRPipeLevel algorithm is as Fig. 10, and consists of the Map(), Reduce(), MultiPipeMap() and MRPipeLevel() functions. The Map() and Reduce() functions are similar to the MRLevel. However, the MRPipeLevel algorithm includes the part to create the sort tree and pipelines on the cube lattice and the MultiPipeMap() function for the multi-pipeline aggregation. First, the MRPipeLevel() function searches the cube lattice by the level, finds the part corresponding to the cuboid's prefix to connect to the pipeline, and other cuboids construct the SortTree through the minimum cost matching. If the whole cube lattice is searched, all the pipelines computable without resorting and the SortTree requiring resort are constructed. First, the Map() and Reduce() are carried out for the SortTree, and then the MultiPipeMap() is conducted for the pipeline.

MRPipeLevel(). MRPipeLevel() functions searches a cube lattice level by level and constructs pipelines of cuboids with the same prefix and sort trees of the other cuboids. For each sort tree, Map() and Reduce() functions are executed. Then MultiPipeMap() function is invoked for the pipelines.

Algorithm   MRPipeLevel
Map()
Input
S is a Sort Tree
Output
⟨*cell, measure*⟩
Description
1 **for each** parent cuboid P in S
2   **for each** child cuboid C in S
3   C(*cell, measure*) ← P(*cell, measure*)
4   ***emit*** ⟨*cell, measure*⟩
Reduce()
Input
⟨*cell, measures* ($m_1, m_2, \cdots, m_n$)⟩
Output
⟨*cell, measure*⟩
Description
1 *measure* = ***function***($m_1, m_2, \cdots, m_n$)
2 ***emit*** ⟨*cell, measure*⟩
MultiPipeMap()
Input
P is Pipelines
⟨*cell, measures* ($m_1, m_2, \cdots, m_n$)⟩
Output
⟨*cell, measure*⟩
Description
1 M = ***function***($m_1, m_2, \cdots, m_n$)
2 **for each** Cuboid C in P
3   **if** *cell* = Cell c in C
4   *measure* = ***function***(*measure* ∪ M)
5   **else**
6   ***emit*** ⟨*cell, measure*⟩
MRPipeLevel()
Description
1 **for each** level k in Cube Lattice
2   Pipelines P = P ∪ **FindPrefixCuboid**(k + 1 → k)
3   SortTree S = S ∪ **MinimumCostMatching**(k + 1 → k)
4 **for each** SortSubtree s_k in S
5   R = R ∪ ***Map***(s_k) ∪ ***Reduce***()
6 R = R ∪ ***MultiPipeMap***(P)
7 **return** R

Fig. 10. The algorithm of MRPipeLevel

MultiPipeMap(). The MultiPipeMap() function processes aggregation for the pipeline introduced in the Sect. 4.1, and could process the multi-pipeline aggregation together. Looking at the algorithm, there could be one or multiple pipe-lines in the pipeline P, and each pipeline's cuboid has a space to store a cell. Comparing each cuboid's cell with the cell coming into as input, measurement is computed to store for the identical cell, the existing cell is emitted for the non-identical cell, and the entered cell is stored.

5   Experiments

5.1   Experimental Setup

In the experiments, we used 1 NameNode, 20 DataNode and total 21 PCs in a cluster. NameNode equipped with Intel Pentium 4 3.0 GHz CPU, 1 GB RAM, and a 400 GB HDD. DataNode equipped with Intel Pentium 4 3.0 GHz CPU, 512 MB RAM, and a 150 GB HDD. The operating system was Ubuntu Linux, the Java was JDK 1.6, and the MapReduce framework was Hadoop 0.20.2. The network speed was 1 Gbps.

In the experiments, we compare our MRPipeLevel algorithm against six algorithms: the naïve MapReduce algorithm as MRNaïve algorithm, MapReduce version of two existing algorithms, GBLP [1] and PipeSort [3, 6], as MRGBLP [16] and MRPipeSort respectively, the MRChildren algorithm extending MRGBLP by calculating multiple chidren cuboids from a parent cuboid in a single MapReduce phase, and the MRLevel algorithm computing every cuboids in a level from their parent cuboids.

5.2   Varying the Data Size

Figure 11 shows the result by varying the data size, where we increase the data size from 20 million to 100 million. As shown in the figure, as the data size increases, the MRNaïve algorithm execution time increases significantly. For all other algorithms, data cube computation time by increasing the data size differences were not significant. However, the MRPipeLevel algorithm execution time showed the fastest rate, MRLevel algorithm showed a similar rate with MRPipeLevel algorithm.

5.3   Varying the Number of Dimensions

Figure 12 shows the elapsed time obtained by varying the number of dimensions. In this experiment, we set the number of tuples to 50 billion, but we increase the number of dimensions from three to nine by one. As shown in Fig. 12, MRPipeLevel algorithm was fastest in all dimensions and MRNaïve algorithm was slowest. It did not work correctly with emitting of too much data in 9-dimensions. MRPipeSort and MRLevel algorithm also did not work for 9-dimensions. MRGBLP algorithm faster than MRNaïve and MRPipeSort faster than MRLevel but MRGBLP, MRPipeLevel were processing normally in 9-dimensions or high-dimensions.

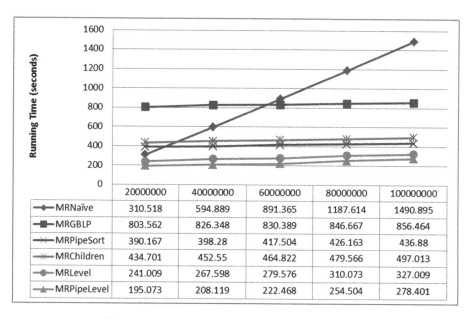

Fig. 11. Elapsed time by varying the number of tuples

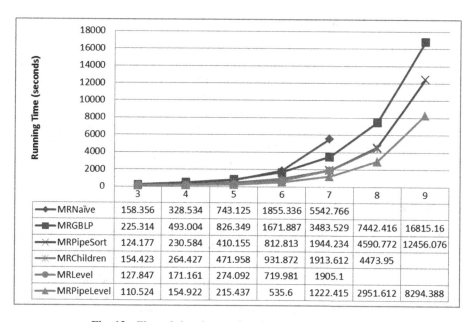

Fig. 12. Elapsed time by varying the number of dimensions

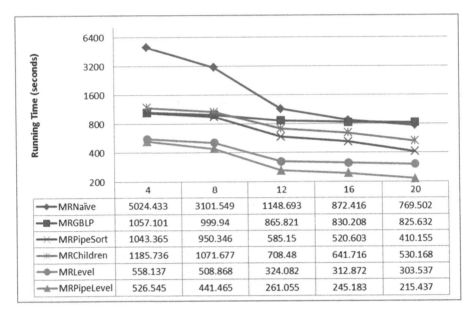

Fig. 13. Elapsed time by varying the number of nodes

5.4   Varying the Number of Nodes

Figure 13 is a comparison between algorithms as increasing the number of nodes. Figure 13 is a result that measures time of each algorithm for 5-dimensional data of 50 billion as increasing the number of nodes from 4 to 20. From the result, it could be identified that the time to compute data cubes is decreased as the number of nodes is increased. MRPipeLevel algorithm was fastest in all nodes and MRNaïve algorithm increases the number of nodes most significant computation time was reduced. Up to 12 nodes, all of the algorithms the computation time was reduced a lot, more than 16 nodes, the computation time can be determined that slightly smaller decrease.

5.5   Performance Analysis of Sort Tree and Pipelines

The MRPipeLevel algorithm includes the part to create the Sort Tree and Pipelines on the cube lattice. The pipelines reuse the sorted cuboid, so the map function is computed without emit data. The sort tree is computed cuboid in parallel by minimizing the cost of the scan. In Fig. 14, the MRPipeLevel algorithm's sort tree and pipelines were examined by (a) varying the data size, (b) varying the number of dimensions (10 M), (c) varying the number of dimensions (50 M), and (d) varying the number of clusters. In Fig. 14(a), pipelines' computation time is nearly constant and sort tree's computation time is increases. In addition, In Fig. 14(b) and Fig. 14(c), pipelines' computation time is smaller than sort tree's computation time by varying the number of dimensions. In Fig. 14(d), pipelines don't use distributed processing because it is computed by scanning only once using sorted cuboid.

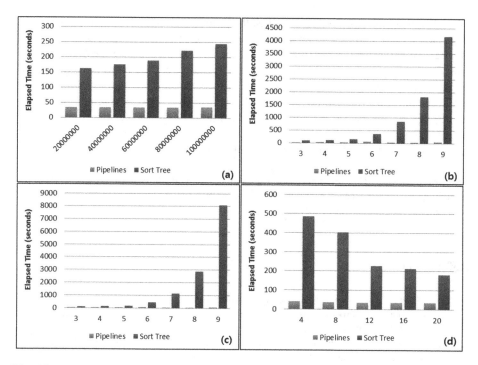

Fig. 14. Elapsed time of sort tree and pipelines by (a) varying the data size, (b) varying the number of dimensions (10 M), (c) varying the number of dimensions (50 M), and (d) varying the number of clusters

6 Conclusion and Future Work

Existing MapReduce data cube algorithm is very fast for low-dimensions (or small size), but it's difficult for high-dimensions (or big size). In addition, it's not suitable for distributed parallel processing of MapReudce. In this paper, we proposed the MRLevel and the MRPipeLevel algorithm to effectively compute the data cube using MapReudce utilizing a large number of PCs to analyze large amounts of data that can be used online. The MRLevel algorithm builds a cube execution tree from a cube lattice to process level by level. The execution tree consists of small parents and children. The MRPipeLevel algorithm extracts the execution plan of sort tree and pipeline on cube lattice structure. Cuboids in sort tree minimize scan cost for each level at a time by MapReduce using distributed parallel computation. Cuboids in pipelines are computed at once using sorted cuboids without emitting the data on each node. Thus, the MRPipeLevel algorithm reduce the computation time of the full cube using a strategy of parallel processing as much data as possible and reducing the data scan.

In this paper, we implement and evaluate the MRPipeLevel algorithm through various experiments. We carry out a diversity of experiments with both low-dimensional or high-dimensional data, and comparative experiments with the MapReduce data cube algorithms which are typical top-down ROLAP data cube computation algorithms. For future works, the proposed method is compared to iceberg cube, closed cube, and bottom-up approach algorithms. Furthermore, we will develop a MapReduce algorithm for suitable iceberg cube or closed cube. We are also going to investigate on approximate query answering technique over data cubes [18] to increase the performance of OLAP query processing. When transmitting data to compute data cubes within the MapReduce framework, it will be a good research to apply privacy-preserving OLAP techniques [19] in order to enhance personal privacy and security. Finally, we will be able to extend data cubes to incorporate various data sources (e.g., XML) and diverse aggregate functions [20].

Acknowledgement. This research work was supported by Basic Science Research Program through the National Research Foundation of Korea (NRF) funded by the Ministry of Education, Science, and Technology (2011-0011824).

References

1. Gray, J., et al.: Data cube: a relational aggregation operator generalizing group-by, cross-tab, and sub-totals. In: Proceedings of Conference on Data Engineering, New Orleans, LA, pp. 152–199, February 1996
2. Harinarayan, V., Rajaraman, A., Ullman, J.D.: Implementing data cubes efficiently. In: Proceedings of International Conference on Management of Data, ACM SIGMOD, Montreal, Canada, pp. 205–216, June 1996
3. Agarwal, S., et al.: On the computation of multidimensional aggregates. In: Proceedings of the 22nd International Conference on Very Large Data Bases, pp. 506–521, September 1996
4. Ross, K.A., Srivastava, D.: Fast computation of sparse datacubes. In: Proceedings of the 23rd International Conference on Very Large Data Bases, pp. 116–125, August 1997
5. Beyer, K., Ramakrishnan, R.: Bottom-up computation of sparse and iceberg cubes. In: Proceedings of International Conference on Management of Data, ACM SIGMOD, Philadelphia, PA, pp. 359–370, June 1999
6. Dehne, F., Eavis, T., Rau-Chaplin, A.: The cgmCUBE project: optimizing parallel data cube generation for ROLAP. Distrib. Parallel Databases **19**(1), 29–62 (2006)
7. Chen, Y., Dehne, F., Eavis, T., Rau-Chaplin, A.: PnP: parallel and external memory iceberg cube computation. In: Proceedings of the International Conference on Data Engineering, Tokyo, Japan, pp. 576–577, April 2005
8. Jin, R., Vaidyanathan, K., Yang, G., Agrawal, G.: Communication and memory optimal parallel data cube construction. Parallel Distrib. Syst. **16**(12), 1105–1119 (2005)
9. Ng, R. T., Wagner, A., and Yin, Y.: Iceberg-cube computation with PC clusters. In: Proceedings of International Conference on Management of Data, ACM SIGMOD, Santa Barbara, CA, pp. 25–36, June 2001
10. Ghemawat, S., Gobioff, H., Leung, S.T.: The Google File System. In: Proceedings of 19th on operating Systems Principles, Bolton Landing, NY, pp. 29–43, December 2003
11. Hadoop. http://hadoop.apache.org/
12. HDFS. http://hadoop.apache.org/hdfs/
13. Dean, J., Ghemawat, S.: MapReduce: simplified data processing on large clusters. Commun. ACM **51**(1), 107–113 (2008)

14. Jinguo, Y., Jianging, X., Pingjian, Z., Hu, C.: A parallel algorithm for closed cube computation. In: Proceedings of 7th International Conference on Computer and Information Science, Portland, OR, pp. 95–99, May 2008

15. Yuxiang, W., Aibo, S., Junzhou, L.: A MapReduceMerge-based data cube construction method." In: Proceedings of 9th International Conference on Grid and Cooperative Computing, Nanjing, China, pp. 1–6, Nov. 2010

16. Lee, S., Moon, Y.-S., Kim, J.: Distributed parallel top-down computation of data cube using MapReduce. In: Proceedings of 3rd International Conference on Emerging Databases, Inchoen, Korea, pp. 303–306, August 2011

17. Nandi, A., Yu, C., Bohannon, P., Ramakrishnan, R.: Distributed cube materialization on holistic measures. In: Proceedings 27th International Conference on Data Engineering, Hannover, Germany, pp. 183–194, April 2011

18. Cuzzocrea, A.: Providing probabilistically-bounded approximate answers to non-holistic aggregate range queries in OLAP. In: Proceedings of 8th International Workshop on Data Warehousing and OLAP, Bremen, Germany, pp. 97–106, November 2005

19. Cuzzocrea, A. Sacca, D.: Balancing accuracy and privacy of OLAP aggregations on data cubes. In: Proceedings of 13th International Workshop on Data Warehousing and OLAP, Toronto, Canada, pp. 93–98, October 2010

20. Cuzzocrea, A., Darmont, J., Mahboubi, H.: Fragmenting very large XML data warehouses via k-means clustering algorithm. Int. J. Bus. Intell. Data Min. **4**(3), 301–328 (2009)

Differentiated Multiple Aggregations in Multidimensional Databases

Ali Hassan[1($\boxtimes$)], Frank Ravat[1], Olivier Teste[2], Ronan Tournier[1],
and Gilles Zurfluh[1]

[1] Université Toulouse 1 Capitole, IRIT (UMR 5505),
118 Route de Narbonne, 31062 Toulouse cedex 9, France
{hassan,ravat,tournier,zurfluh}@irit.fr
[2] Université Toulouse 3 Paul Sabatier, IRIT (UMR 5505),
118 Route de Narbonne, 31062 Toulouse cedex 9, France
teste@irit.fr

Abstract. Many models have been proposed for modeling multidimensional data warehouses and most consider a same function to determine how measure values are aggregated according to different data detail levels. We provide a conceptual model that supports (1) multiple aggregations, associating to the same measure a different aggregation function according to analysis axes or hierarchies, and (2) differentiated aggregation, allowing specific aggregations at each detail level. Our model is based on a graphical formalism that allows controlling the validity of aggregation functions (distributive, algebraic or holistic). We also show how conceptual modeling can be used, in an R-OLAP environment, for building lattices of pre-computed aggregates.

Keywords: Data warehouse · Conceptual modeling · Aggregate lattice · Multiple aggregations · Aggregation functions

1 Introduction

Decision support systems, such as data warehouses, have shown their ability to integrate large volumes of data by supporting effectively the analysis of stored data. These decision support systems are elaborated from data sources, usually the operational system of an organization; the data identified in the relevant sources are extracted, transformed and loaded [26] in a storage area called a data warehouse. To allow efficient querying and analysis of the data, specific data organization techniques have been developed using multidimensional databases (MDB) [3, 13]. This type of modeling considers the analyzed data from analysis indicators (i.e. measures grouped into facts) as points in a multidimensional space, forming a data cube [8]. Each dimension having various granularity/detail levels. Decision makers visualize extracts of data cubes, usually a two-dimensional slice of a cube. From this structure, called a multidimensional table (MT) [9], the decision maker can interact through manipulation operations [22]. The most emblematic operations are drilling operations which change the granularity level of the analyzed data and rotation operations which change the slice

© Springer-Verlag Berlin Heidelberg 2015
A. Hameurlain et al. (Eds.): TLDKS XXI, LNCS 9260, pp. 20–47, 2015.
DOI: 10.1007/978-3-662-47804-2_2

of the cube. These operations are the most popular ones used for On-Line Analytic Processing (OLAP).

This environment provides a suitable analysis framework for decision makers; however, the imposed structure may be imperfect. In particular, a classical MDB supports only the calculation of a measure made by the same aggregation function while performing drilling or rotating operations (i.e. changing the analyzed slice of the cube). For example, if we consider sales amounts, these can be calculated as the sum of the products sold by cities and years (top part of Fig. 1). When drilling from cities to countries, the new amounts are calculated using the same aggregation function (SUM in the bottom part of Fig. 1). When the user wishes to change the aggregation function between two slices of the manipulated cube, the classical BDM no longer guarantees the validity of the calculated data, or even worse: does not support this type of manipulation.

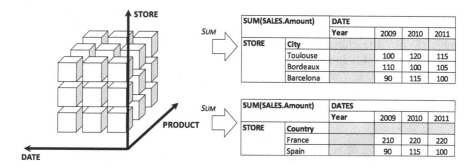

Fig. 1. Uniform aggregation in slices of a cube

This paper aims at allowing non-uniform aggregations during user manipulations. To ensure the validity of such aggregations, we define **differentiated multiple aggregations**. Our proposal aims at developing a multidimensional model flexible enough for designing cubes with aggregation functions according to different levels.

1.1 Case Study

The case study concerns a diploma delivery jury. Here, decision makers (jury members) deliver diplomas by analyzing the marks (average, maximum, minimum) of students and their rate of absenteeism.

Students are split into groups and the academic year has two semesters. Each semester consists of Teaching Units (TU) and each TU is composed of several courses. Each course is associated with a coefficient that represents the importance of the course in the TU. We must take into account this coefficient to calculate the mark of the EU, which itself is linked to an ECTS (European Credit Transfer System) used to calculate the mark of semester. Each semester has the same amount of ECTS. In addition to the courses and students, analysts can analyze marks and absenteeism rates according to the dates (academic years).

Analysts may wish to observe absenteeism in two different ways:

- The first, called **simple**, is to calculate the percentage of absenteeism without distinction between different courses or TUs.
- The second, called **weighted**, uses the same coefficients (used for calculating the marks of TUs and semesters) to calculate absenteeism rates.

An MDB is implemented using extracting, transforming processes and loading data from the operational system, which we will not detail in this article. Figure 2 shows the conceptual star schema [7, 21] of the MDB of our case study. This MDB analyzes the measures (average marks 'Avg_Mark', maximum marks 'Max_Mark', minimum marks 'Min_Mark' and absenteeism rates 'Rate_Abs') by 'Courses', 'Students' and 'Dates' (dimensions).

The dimension 'Courses' has two hierarchies 'HCourse_Simple' and 'HCourse_weighted'. Each hierarchy corresponds to a way to analyze the absenteeism rate (simple and weighted). The other measures ('Avg_Mark', 'Max_Mark' and 'Min_Mark') are analyzed in the same way on the two hierarchies. A course is characterized by a course number (C_Id), a teaching unit number (TU_Id) and a semester. Each student has a student number (S_Id) and a group number (G_Id). Academic years 'Academic_year' of the dimension 'Dates' are aggregated by periods of five years 'period-5' and periods of ten years 'period-10'.

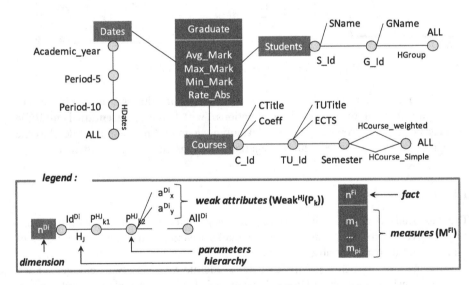

Fig. 2. The MDB of the diploma delivery case study

1.2   Illustration of the Problem

This schema allows getting average marks by courses and by students (Fig. 3). Obtaining the average mark by TU in this multidimensional environment requires aggregating the average marks by courses in accordance with the function associated

AVG(Graduate.Mark)		Courses			
		Semester	S1		
		TU_Id	U1		U2
		C_Id	M1	M2	M3
Students	S_Id (SName)				
	St1 (Tom)		14	10	12
	St2 (Sara)		8	10	9

Fig. 3. A MT visualizing the student's average marks by course

with the measure Mark (AVG). But this operation gives a result that does not correspond to examination modalities: an average mark by TU should be calculated from the course marks and taking into account the coefficient of each course (Eq. 1). Similarly, for average marks by semester, the ECTS of each TU (Eq. 2) has to be taken into account. However, to calculate the general average mark for each student, one must calculate the average of the TU marks (Eq. 3).

$$AVG\_TU = \frac{\sum Mark \times Coeff}{\sum Coeff} \tag{1}$$

$$AVG\_Semester = \frac{\sum AVG\_TU \times ECTS}{\sum ECTS} = \frac{\sum \left(\frac{\sum Mark \times Coeff}{\sum Coeff} \right) \times ECTS}{\sum ECTS} \tag{2}$$

$$AVG\_ALL = AVG(AVG\_Semester) \tag{3}$$

Therefore, classical approaches that consider a single aggregation function for all modeled aggregation levels in the star schema suffer from several limits:

- **Variability of the aggregation function.** Traditionally, models do not allow the use of aggregation functions that vary along dimensions or hierarchical levels. In our example, the aggregation function changes between the levels C_Id (courses), TU_Id (teaching units) and the semester level.
- **Shortcomings of basic functions.** When aggregating data across hierarchical levels, in our example, we use non-standard aggregation functions which use complementary data other than measure values (i.e. coefficients *Coeff*, weights *ECTS*).
- **Aggregation constraints.** The way to make the calculation of aggregation functions may be constrained. In our example, as shown in (Eq. 2), the average per semester cannot be obtained directly from the marks per courses. It is necessarily calculated from the averages per TUs. Similarly, the general average is calculated from the averages per semesters (Eq. 3).

The objective of this paper is to propose a multidimensional model sufficiently expressive to support these various aggregations. Then, we study the impacts of this conceptual model on the lattice of pre-aggregates [8] at the logical level.

In previous work [11], we detailed our conceptual model and presented simply the logical model. Here, we:

- Extend the conceptual model with a new type of aggregation (hierarchical);
- Revisit the execution order mechanism in order to be more expressive;
- Detail the logical model;
- Implement our prototype to study the consequences on lattice reductions.

The rest of this paper is organized as follows: Sect. 2 reviews related work. Section 3 defines our conceptual multidimensional model: classical concepts, extensions for differentiated multiple aggregations and associated graphical formalisms. Section 4 shows the logical R-OLAP model of our star schema and its optimization relations. We detail our prototype and experiments in Sect. 5 and the last section concludes this work and states some future work.

2   Related Work

There are typically two approaches for modeling multidimensional databases. The first is based on the data cube (or hypercube) metaphor according to which the MDB is represented by cubes. The second is known as multidimensional modeling, where the MDB is described by a star schema or a constellation [13]. Our work falls in the second category. A cube is based on an equivocal separation between the structure elements and the values [24]: modeling analysis axes is not very expressive especially due to the difficulty for representing the hierarchical organization of the data. It is also limited for representing constellations of facts with shared dimensions.

Several surveys of the domain [3, 17, 25] and comparative studies [1, 2, 7, 9, 14, 16, 18–23, 27] are available in the scientific literature. One, [17], deals with problems related to complex structures such as non-strict, roll-up incomplete and drill-down incomplete hierarchies. We don't address this kind of problem. We focus on the problem of using several aggregation functions during an analysis.

Most of the existing proposals consider that a measure is associated with only one aggregation function for all aggregation levels. This function calculates the same aggregation for all combinations of all modeled parameters.

The treatment of aggregation of measures in the multidimensional space has evolved (Table 1). Two contributions [9, 27] do not specify aggregation functions at the measure level; however, they leave the possibility to use several aggregation functions for each measure during OLAP analyses. This provides great flexibility, but allows the user to do errors by using inappropriate aggregation functions. In addition, one advantage of specifying the aggregation functions in the conceptual model is to use them for the cube computation, i.e. for the pre-computation of the aggregates. In [19, 21, 23], the authors, in theirs conceptual models, can link to a single measure a set of functions which includes only valid functions. However, in these three papers, the same function will be used with all the dimensions and all aggregation levels.

In [4, 6], the authors assume that the aggregation function is determined for a measure in the analysis queries. This function can change from one query to another one while concerning the same measure. But in each query, the aggregation function will be used uniformly over all the dimensions involved in the analysis. In [5], although the authors store multiple aggregations data in a hierarchical organization according to the time granularities, they use the same function for all granularities.

The YAM^2 model [1] and the work presented in [7] support a different aggregation function with each dimension. However, these models do not support function change neither between hierarchies nor within the hierarchical levels. This limit has been lifted by the aggregation model of [20] which allows associating an aggregation function to each dimension or each hierarchy or sub-hierarchy, but the model considers only standard functions (SUM, AVG, MIN, MAX and COUNT). In [2] the authors overcome this limit. However, these last two papers [2, 20] suffer from a limitation: the authors do not consider the case where aggregation functions are non-commutative (for example, average and weighted average).

Regarding commercial tools, "Business Objects" uses a single aggregation function for each measure. By contrast, "Microsoft Analysis Services" offers the possibility that a "custom rollup" can be applied in a hierarchy in several ways [10]:

- By using unary operators to solve the aggregation problem over a particular type of hierarchy (parent-child attributes hierarchy). These hierarchies are built from a single attribute with a reflexive join relationship on the attribute itself (i.e. technically a join on the dimension table itself).
- By using MDX scripts, either directly or by using the attribute property "CustomRollupColumn" which indicates a column where MDX scripts are stored.

These two ways concern aggregation functions but it is not related to a specific dimension or an aggregation level. It is related to a member (an instance) of an aggregation level in a hierarchy (i.e. a line in the dimension table). Therefore, applying this "custom rollup" to a single aggregation level requires repeating it for all the instances of that level. This poses a storage problem and reduces performance [10]. Moreover, binding a "custom rollup" with a specific instance can cause difficulties when updating data.

The MDX language allows the possibility for building data sets (that will be aggregated by aggregation functions) using functions: PeriodsToDate, YTD, QTD, MTD, Crossjoin, Cousin, Descendants, Children, Hierarchize, and Members. However, this possibility is not related to our problem: changing the aggregation function according to a considered analysis dimension or hierarchy or level.

The above was about how the integration of aggregation functions within the multidimensional model. But, there is another point that should be taken into consideration; it is the aggregation functions itself. Aggregation functions are classified:

- From an aggregation mechanism point of view, aggregation functions belong to three different categories [8]: The first corresponds to **distributive** functions that calculate aggregated values of the selected granularity level from the values already aggregated at the lower level (e.g. yearly amounts can be calculated by summing monthly values). The second corresponds to **algebraic** functions that calculate aggregated values from stored intermediate results (for example, the average of an amount per year can be calculated from the sum of the amounts and counting the occurrences from a month level). Finally, the third corresponds to **holistic** functions that cannot be calculated from intermediate results. In this case, aggregated values must be calculated from the elementary values of the lowest granularity level (e.g. RANK).

- From a summerizability point of view, aggregation functions are classified in two groups [1]: (1) "Transitive" that guarantees summerizability, (2) "Non-Transitive" which implies that aggregations must always be calculated from the base level.
- From a measure (data) point of view, aggregation functions are of three types [19]: (1) for additive data, (2) for snapshot data that can be used for average calculations, (3) for constant data, i.e. data that can only be counted.

All these proposals as well as aggregation functions classifications assume that the measure aggregation can be calculated from the base level. Our goal is to add the means to consider the opposite case (when the measure cannot be aggregated from base level).

Table 1 shows in the column 'General' if existing proposals integrate the aggregation functions in the OLAP process (during the interrogation) or in the multidimensional model. It also shows if these proposals offer the possibility to change the aggregation functions with the dimensions, hierarchies, and granularity levels (columns 'Dimension', 'Hierarchy' and 'Granularity Level'). In addition, it presents if the proposals treat the case of non-commutative functions (column 'Non-commutative') or the case of constraint aggregations, i.e. when the measure must be calculated from a different level of the base level (column 'Constraint aggregation').

Table 1. Synthesis of works on multidimensional aggregations

	General	Multiple aggregation		Differentiated aggregation	Non-commutative	Constraint aggregation
		Dimension	Hierarchy	Granularity Level		
Gyssens, 1997 [9]	OLAP	–	–	–	–	–
Golfarelli, 1998 [7]	Model	✓	–	–	–	–
Vassiliadis, 2000 [27]	OLAP	–	–	–	–	–
Pedersen, 2001 [19]	Model	–	–	–	–	–
Cuzzocrea, 2004, 2005, 2010 [4–6]	Query	–	–	–	–	–
Abelló, 2006 [1]	Model	✓	–	–	✓	–
Ravat, 2007 [21]	Model	–	–	–	–	–
Silva, 2008 [23]	Model	–	–	–	–	–
Prat, 2010 [20]	Model	✓	✓	✓	–	–
Boulil, 2011 [2]	Model	✓	✓	✓	–	–
Business Objects	Model	–	–	–	–	–
Microsoft Analysis Services	Model	–	–	✓ (Instance)	✓	–

Using this table, we see that the ability to change the aggregation function with the dimensions, hierarchies, and aggregation levels had been treated [2, 20], but partially because these proposals do not take into account the non-commutative functions. Moreover, the works taking into account the non-commutative functions support only changing aggregation functions with the dimensions [1] or the instances of aggregation levels [10].

To our knowledge, all existing proposals assume that it is possible to calculate the measure aggregation from base levels. We propose to add a way to handle the case where the measure must be calculated from a specific level other than the base level (column 'Constraint aggregation') using *aggregation constraints*.

Our aim is to remove these limits by designing a conceptual multidimensional model for representing differentiated multiple multidimensional aggregates. By *multiple* we mean that the same measure can be aggregated by several aggregation functions according to hierarchies or analysis axes and by *differentiated* we mean that these aggregations may vary, depending on the chosen aggregation level.

Our proposal conceptual can be implemented as a layer on top of an OLAP engine, so that it can take into account these multiple and differentiated aggregation functions. In this paper we have declined this conceptual model in a relational framework.

3   Conceptual Data Model

3.1   Classical Concepts

Let us define $\mathcal{N}$, F and D such as: $\mathcal{N} = \{n_1, n_2, \ldots \}$ a finite set of non-redundant names; $F = \{F_1, \ldots, F_n\}$ is a finite set of facts, $n \geq 1$; and $D = \{D_1, \ldots, D_m\}$ is a finite set of dimensions, $m \geq 2$.

Definition 1. A *fact*, denoted F_i, $\forall\ i \in [1..n]$, is defined by (n^{Fi}, M^{Fi}), where:

- $n^{Fi} \in \mathcal{N}$ is the name that identifies the fact,
- $M^{Fi} = \{m_1, \ldots, m_{pi}\}$ is a set of *measures* or indicators.

We define the measure set as

$$M = \cup_{i=1}^{n} M^{F_i}$$

Definition 2. A *dimension*, denoted D_i, $\forall\ i \in [1..m]$, is defined by (n^{Di}, A^{Di}, H^{Di}), where:

- $n^{Di} \in \mathcal{N}$ is the name that identifies the dimension,
- $A^{Di} = \left\{a_1^{D_i}, \ldots, a_{r_i}^{D_i}\right\} \cup \{\ Id^{Di}, All^{Di}\}$ is the set of the *attributes of the dimension*,
- $H^{Di} = \left\{H_1^{D_i}, \ldots, H_{s_i}^{D_i}\right\}$ is a set of *hierarchies*.

Hierarchies organize the attributes of a dimension, from the finest graduation (root parameter Id^{Di}) to the most general graduation (extremity parameter, All^{Di}). Thus a hierarchy defines the valid navigation paths on an analysis axis.

We define the attribute set and the hierarchy set respectively as

$$A = \cup_{i=1}^{m} A^{D_i} \text{ and } H = \cup_{i=1}^{m} H^{D_i}$$

Definition 3. A *hierarchy*, denoted H_j (abusive notation of $H_j^{D_i}$, $\forall i \in [1..m]$, $\forall j \in [1..s_i]$) is defined by $(n^{Hj}, P^{Hj}, \prec^{Hj}, \text{Weak}^{Hj})$, where:

- $n^{Hj} \in N$ is the name that identifies the hierarchy,
- $P^{Hj} = \left\{ p_1^{H_j}, \ldots, p_{q_j}^{H_j} \right\}$ is a set of attributes called *parameters*, $P^{Hj} \subseteq A^{Di}$,
- $\prec^{Hj} = \left\{ \left(p_x^{H_j}, p_y^{H_j} \right) | p_x^{H_j} \in P^{Hj} \wedge p_y^{H_j} \in P^{Hj} \right\}$ is an antisymmetric and transitive binary relation between parameters. Remember that the antisymmetry means that $\left(p_{k_1}^{H_j} \prec^{H_j} p_{k2}^{H_j} \right) \wedge \left(p_{k_2}^{H_j} \prec^{H_j} p_{k1}^{H_j} \right) \Rightarrow p_{k1}^{H_j} = p_{k2}^{H_j}$ while the transitivity means that $\left(p_{k_1}^{H_j} \prec^{H_j} p_{k2}^{H_j} \right) \wedge \left(p_{k_2}^{H_j} \prec^{H_j} p_{k3}^{H_j} \right) \Rightarrow p_{k1}^{H_j} \prec^{H_j} p_{k3}^{H_j}$.
- $\text{Weak}^{Hj} : P^{Hj} \rightarrow 2^{A^{D_i} \setminus P^{H_j}}$ is an application that associates to each parameter a set of dimension attributes, called *weak attributes* (2^N represents the power set of N).

We define parameter sets as

$$P^{D_i} = \cup_{j=1}^{s_i} P^{H_j} \text{ and } P = \cup_{i=1}^{m} P^{D_i} = \cup_{i=1}^{m} \cup_{j=1}^{s_i} P^{H_j}$$

Lemma 1. For each dimension D_i, a *root parameter*, denoted $Id^{Di} \in P^{Di}$, exists. It is defined as follows: $\forall j \in [1..s_i], \forall p_k^{H_j} \in p^{Di}, Id^{Di} \neq p_k^{H_j} | Id^{Di} \prec^{H_j} p_k^{H_j}$.

Lemma 2. For each dimension D_i, a *extremity parameter*, denoted $All^{Di} \in P^{Di}$, exists. It is defined as follows: $\forall j \in [1..s_i], \forall p_k^{H_j} \in p^{Di}, All^{Di} \neq p_k^{H_j} | p_k^{H_j} \prec^{H_j} All^{Di}$.

Lemma 3. For each dimension D_i, all its attributes are exclusively either parameters or weak attributes, $P^{Di} \cap W^{Di} = \varnothing$ and $P^{Di} \cup W^{Di} = A^{Di}$.

3.2   Extensions for Differentiated Multiple Aggregations

We enrich the multidimensional model for specifying how the aggregations calculations are performed during OLAP analysis. This corresponds to three extensions:

- The first extension concerns the aggregation process which allows using several aggregation functions for the same measure:

  - **Differentiated aggregation.** It consists in aggregating measure values between two parameters (aggregation levels) of a hierarchy. The aggregation function is associated with one measure and one parameter. This kind of aggregation allows a specific aggregation over each level of granularity.
  - **Multiple hierarchical aggregation.** It is used to aggregate the measure values between all the parameters over a hierarchy. This is a simplified representation

instead of a repeated use of the same differentiated function over several levels of granularity. It is important to note that several aggregation functions can be associated to a same measure; one for each hierarchy.

- **Multiple dimensional aggregation**. It consists in aggregating measure values using different aggregation functions depending on the used dimension. Similarly to multiple hierarchical aggregation, multiple dimensional aggregation is a simplified representation instead of a repeated use of the same multiple hierarchical aggregation over several hierarchies. The same aggregation is performed over each level of granularity of a dimension. The function is associated with one measure and a dimension.

- **General aggregation**. This function is associated only with a measure without taking into account neither parameter nor hierarchy nor dimension. This is a simplified representation instead of a repeated use of the same multiple dimensional function over several dimensions. This is equivalent to aggregation functions in classical models.

- The second extension concerns the **execution order** for handling the case of non commutative aggregation functions. It is possible to have different aggregation functions during an analysis. These functions are generally not commutative. Therefore, it is necessary to plan in the MDB an execution order when using the functions between the different dimensions.

- The third extension concerns **aggregation constraints** which aim at handling the case where the measure cannot be calculated from the base level. All aggregations are not carried out uniformly using systematically all lower hierarchical levels (contrarily to the aggregation process designed in classical multidimensional models). Therefore, we introduce a constraint mechanism on the aggregation process to indicate the valid aggregation level that allows obtaining the upper level.

Let $F = \{f_1, f_2, \ldots\}$ be a finite set of aggregation functions.

Definition 4. A *multidimensional schema*, denoted S, is defined by (F, D, Star, Aggregate), where:

- $F = \{F_1, \ldots, F_n\}$ is the set of facts, if $|F| = 1$ then the multidimensional schema is called a *star schema* while if $|F| > 1$ it is a *constellation schema*,
- $D = \{D_1, \ldots, D_m\}$ is the set of dimensions,
- Star: $F \rightarrow 2^D$ is a function that associates each fact to a set of dimensions according to which it can be analyzed.
- Aggregate: $M \rightarrow 2^{N* \times \mathcal{F} \times 2^D \times 2^H \times 2^P \times N^-}$ associates each measure to a set of aggregation functions. Aggregate defines the different types of aggregation functions supported by our model:

  - If 2^D, 2^H and 2^P are not used ($2^D = \varnothing$, $2^H = \varnothing$ and $2^P = \varnothing$) then the function is a *general aggregation* function.
  - If 2^H and 2^P are not used ($2^D \neq \varnothing$, $2^H = \varnothing$ and $2^P = \varnothing$) then the function is a *multiple dimensional aggregation* function. Here, the function aggregates the measure over the entire considered dimension.

- If 2^P only is not used ($2^D \neq \emptyset$, $2^H \neq \emptyset$ and $2^P = \emptyset$) then the function is a *multiple hierarchical aggregation* function. Here, the function aggregates the measure over the entire considered hierarchy.
- If $2^D \neq \emptyset$, $2^H \neq \emptyset$ and $2^P \neq \emptyset$ then the function is a *differentiated aggregation* function. Here, the function aggregates the measure between a considered parameter and the parameter directly above it in the same hierarchy.

$\mathbb{N}^*$ binds to each function an execution order. The aggregation function with the smallest order is the highest priority. If the aggregation functions are commutative, then both functions will have the same order. Choosing a valid order depends on the requirements of the user. It may differ from one case to another, even if the functions are the same in both cases.

$\mathbb{N}^-$ is to constraint aggregations by indicating a specific level from which the considered aggregation must be calculated. An unconstrained aggregation will be associated with 0 while a constrained aggregation will be associated with a negative value to force the calculation from a chosen level lower than the considered level.

Lemma 4. Aggregation functions ensure the full *coverage* of multidimensional schemas. Thus there does not exist any parameter (i.e. aggregation levels) for which the aggregation function to be applied is unknown.

$$\forall i \in [1..n], \forall m_k \in M^{Fi}, \exists f \in \mathcal{F}, \exists x_1 \in \mathbb{N}^*, \exists x_2 \in \mathbb{N}^-,$$

$$\left\{ \begin{array}{c} \left| (x_1 f, \{\}, \{\}, \{\}, x_2) \in \textit{Aggregate}(m_k) \right. \\[6pt] \forall D_j \in \textit{Star}(F_i) \left| (x_1 f, \{D_j\}, \{\}, \{\}, x_2) \in \textit{Aggregate}(m_k) \right. \\[6pt] \forall H_s \in H^j \left| (x_1 f, \{D_j\}, \{H_s\}, \{\}, x_2) \in \textit{Aggregate}(m_k) \right. \\[6pt] \forall P_q \in P^s \backslash \{All^j\} \left| (x_1 f, \{D_j\}, \{H_s\}, \{P_q\}, x_2) \in \textit{Aggregate}(m_k) \right. \end{array} \right.$$

Less formally, the *coverage* of the schema is carried out in several ways:

- By using a general aggregation function,
- By using a multiple dimensional aggregation function for each dimension,
- By using a multiple hierarchical aggregation function for each hierarchy,
- By using a differentiated aggregation function for each aggregation level,
- By combining multiple aggregation functions with differentiated ones. Each dimension or hierarchy having no multiple function must have a differentiated function for each aggregation level (i.e. parameter).

3.3 Formalisms

Textual Formalisms. The example of the diploma delivery illustrated in case study, is defined formally by (F, D, Star, Order, Aggregate) where:

- $F = \{F_{Graduate}\}$, where the fact is defined by $F_{Graduate}$ = (*'Graduate'*, {Avg_Mark, Max_Mark, Min_Mark, Rate_Abs }).
- $D = \{D_{Courses}, D_{Students}, D_{Dates}\}$, where the dimensions are defined by:
  - $D_{Courses}$ = (*'Courses'*, {a_{C_Id}, a_{Coeff}, a_{CTitle}, a_{TU_Id}, a_{ECTS}, $a_{TUTitle}$, $a_{Semester}$, $ALL^{DCourses}$}, {$H_{HCourse\_Simple}$, $H_{HCourse\_weighted}$}) with
    - $H_{HCourse\_Simple}$ = (*'HCourse_Simple'*, {a_{C_Id}, a_{TU_Id}, $a_{Semester}$, $ALL^{DCourses}$}, {(a_{C_Id}, a_{TU_Id}), (a_{TU_Id}, $a_{Semester}$), ($a_{Semester}$, $ALL^{DCourses}$)}, {(a_{C_Id}, {a_{Coeff}, a_{CTitle}}), (a_{TU_Id}, {a_{ECTS}, $a_{TUTitle}$})}),
    - $H_{HCourse\_Simple}$ = (*'HCourse_Simple'*, {a_{C_Id}, a_{TU_Id}, $a_{Semester}$, $ALL^{DCourses}$}, {(a_{C_Id}, a_{TU_Id}), (a_{TU_Id}, $a_{Semester}$), ($a_{Semester}$, $ALL^{DCourses}$)}, {(a_{C_Id}, {a_{Coeff}, a_{CTitle}}), (a_{TU_Id}, {a_{ECTS}, $a_{TUTitle}$})}).

  - $D_{Students}$ = (*'Students'*, {a_{S_Id}, a_{SName}, a_{G_Id}, a_{GName}, $ALL^{DStudents}$}, {H_{HGroup}}) with
    - H_{HGroup} = (*'HGroup'*, {a_{S_Id}, a_{G_Id}, $ALL^{DStudents}$}, {(a_{S_Id}, a_{G_Id}), (a_{G_Id}, $ALL^{DStudents}$)}, {(a_{S_Id}, {a_{SName}}), (a_{G_Id}, {a_{GName}})}).

  - D_{Dates} = (*'Dates'*, {$a_{Academic\_year}$, $a_{Period-5}$, $a_{Period-10}$, ALL^{DDates}}, {H_{HDates}}) with
    - H_{HDates} = (*'HDates'*, {$a_{Academic\_year}$, $a_{Period-5}$, $a_{Period-10}$, ALL^{DDates}}, {($a_{Academic\_year}$, $a_{Period-5}$), ($a_{Period-5}$, $a_{Period-10}$), ($a_{Period-10}$, ALL^{DDates})}).

- Star : $F \rightarrow 2^{D}$ |
  Star($F_{Graduate}$) = {$D_{Courses}$, $D_{Students}$, D_{Dates} }
- Aggregate : $M \rightarrow 2^{N* \times \mathcal{F} \times 2^D \times 2^H \times 2^P \times N^-}$ |

Aggregate (Avg_Mark) = {(2, AVG(Avg_Mark), {}, {}, {}, 0),[1]
(1, AVG_W(Avg_Mark, Coeff), {Courses}, {HCourse_weighted}, {C_Id}, 0),
(1, AVG_W(Avg_Mark, ECTS), {Courses}, {HCourse_weighted}, {TU_Id},-1),[2]
(1, AVG(Avg_Mark), {Courses}, {HCourse_weighted}, {Semester}, -1),
(1, AVG_W(Avg_Mark, Coeff), {Courses}, {HCourse_Simple}, {C_Id}, 0),
(1, AVG_W(Avg_Mark, ECTS), {Courses}, {HCourse_Simple}, {TU_Id}, -1),
(1, AVG(Avg_Mark), {Courses}, {HCourse_Simple}, {Semester}, -1)}
Aggregate (Max_Mark) = {(2, MAX(Avg_Mark), {}, {}, {}, 0),
(1, AVG_W(Avg_Mark, Coeff), {Courses}, {HCourse_weighted}, {C_Id}, 0),

---

[1] Note that there is no constraint on the aggregation.

[2] The aggregated values are computed from the values at the level directly below the one considered.

(1, AVG_W(Avg_Mark, ECTS), {Courses}, {HCourse_weighted}, {TU_Id},-1),
(1, AVG(Avg_Mark), {Courses}, {HCourse_weighted}, {Semester}, -1),
(1, AVG_W(Avg_Mark, Coeff), {Courses}, {HCourse_Simple}, {C_Id}, 0),
(1, AVG_W(Avg_Mark, ECTS), {Courses}, {HCourse_Simple}, {TU_Id}, -1),
(1, AVG(Avg_Mark), {Courses}, {HCourse_Simple}, {Semester}, -1)}

Aggregate (Rate_Abs) = {(2, AVG(Rate_Abs), {}, {}, {}, 0),
(1, AVG_W(Rate_Abs, Coeff), {Courses}, {HCourse_weighted}, {C_Id}, 0),
(1, AVG_W(Rate_Abs, ECTS), {Courses}, {HCourse_weighted}, {TU_Id},-1),
(1, AVG(Rate_Abs), {Courses}, {HCourse_weighted}, {Semester}, -1)}

Aggregate(Min_Mark) is identical to Aggregate(Max_Note) except that it uses the MIN function instead of MAX.

The function Avg_W(X, Y) takes as input tow numerical parameters. It returns the average of values of X weighted by Y. In other words, the weighted average:

$$Avg\_W(X, Y) = \frac{\Sigma(X \times Y)}{\Sigma Y}$$

Regarding the measures 'Avg_Mark', 'Max_Mark' and 'Min_Mark', it is aggregated in an identical way on the two hierarchies of the dimension 'Courses'. Moreover, the aggregation of the measures 'Max_Mark' and 'Min_Mark' is based on the aggregation of 'Avg_Mark'. This clearly appears through the use of the measure 'Avg_Mark' in aggregation functions of 'Max_Mark' and 'Min_Mark'. For the maximum mark 'Max_Mark' of a course or a teaching unit for a group of students, we must first calculate the mark 'Avg_Mark' of this course or TU for each student, and then we determine from the obtained marks, the maximum mark.

If we analyze for example the average marks 'Avg_Mark' using the dimensions 'Dates' and 'Students', the decisional system must use the general function 'AVG (Avg_Mark)' to aggregate the measure values because there is no other specific function for these dimensions. If we analyze using the dimension 'Courses', the system uses on each aggregation level a different differentiated aggregation function. Aggregation is done using the level directly below (AVG_W to aggregate the 'TU_Id' and 'Semester' levels using the 'C_Id' and 'TU_Id' levels respectively and AVG for 'ALL' level using the 'Semester' level). Furthermore, if we analyze data using two or more dimensions then functions over the dimension 'Courses' are a priority; that means that we must apply it before the other functions.

Graphical Formalisms. Associated with the formal definitions, we introduce a two-level graphical formalism for easing the understanding of the MDB schema:

- **Structural Schema.** The structural schema is used to display globally the multi-dimensional elements (facts, dimensions and hierarchies) hiding aggregation mechanisms. This global view (see Fig. 2) is defined by the function *Star*. The graphical formalism is based on [7, 22].
- **Aggregation schema.** For each measure $m_k \in M^{Fi}$, an aggregation schema is obtained using the function *Aggregate*. This schema details the aggregation

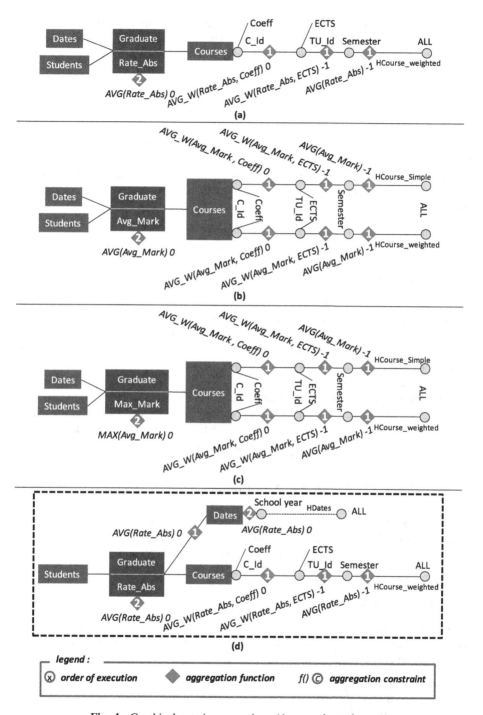

Fig. 4. Graphical notation extensions (Aggregation schemas)

mechanisms involved in the selected measure analysis (multiple, differentiated and general aggregations, constraints of aggregation and execution order) but shows simply the structural elements directly related to the measure. This schema is an extension of our previous work [11].

Figure 4 illustrates three aggregation schemes (a, b, c) corresponding to the measures 'Rate_Abs', 'Avg_Mark' and 'Max_Mark' (we do not present the measure 'Min_Mark'). As shown in Fig. 4, the hierarchies are presented in split version, unlike the structural schema (Fig. 2) where it is presented in compact version, e.g. hierarchies 'HCourse_weighted', 'HCourse_Simple' in Fig. 4 (b and c).

The aggregation functions are modeled by diamonds. Each diamond also indicates the execution order and the possible aggregation constraint. The positions of the diamonds depend on the type of function:

- A general function is represented by a diamond on the fact,
- A multiple dimensional function is on the edge connecting facts to dimensions,
- A multiple hierarchical function is represented on the bottom of the hierarchy,
- A differentiated aggregation function is a label on the edge linking two parameters.

Figure 4 (d) presents multiple aggregation (dimensional and hierarchical) functions and commutativity in the execution order. We assume that there is a multiple dimensional function AVG(Rate_Abs) on the dimension 'Dates'. This function is commutative with the functions of the dimension 'Courses'. We assume also that there is a multiple hierarchical function on the hierarchy 'HDates'. This function is commutative with the general function.

Aggregation with a constraint assigned to -1 is calculated from the directly lower level. E.g. the average mark 'Avg_Mark' by semester is calculated from average marks by UEs. In case, we would have chosen to calculate this average by semester from the marks by courses, the constraint would be assigned to -2.

4   Relational-OLAP (R-OLAP) Logical Model

Current multidimensional schema implementations use mainly the relational approach R-OLAP [13]. This approach has many advantages such as reusing proven data management mechanisms and the ability to manage very large volumes of data.

4.1   R-OLAP Star

In this relational context, the conceptual multidimensional structures (facts and dimensions) are translated into relations [13]. Applied to our example, the R-OLAP schema is the following:

```
COURSES (C_Id, Coeff,CTitle,TU_Id,ECTS,TUTitle,Semester)
STUDENTS (S_Id, SName, G_Id, GName)
DATES (Academic_year, Period-5, Period-10)
GRADUATE (C_Id#,S_Id#,Academic_year#, Avg_Mark, Rate_Abs)
```

According to aggregation functions for the maximum and minimum marks 'Max_Mark', 'Min_Mark' (cf. Fig. 4); these measures are calculated from the average mark measure 'Avg_Mark'. So their values can be obtained directly from those of the measure 'Avg_Mark' without storing them in the relational table corresponding to the fact of model.

The aggregation functions are stored in the database. We use a meta-schema (not detailed here, for more information see [12]) to describe the multidimensional schema (facts, dimensions and hierarchies) corresponding to the R-OLAP relations that store the analysis data. It also describes the different aggregation functions and the possible aggregation constraints.

4.2   Optimized Star

Conceptual modeling allows structuring hierarchically the analysis axis (dimension) graduations (parameters). These hierarchies are exploited for pre-computing the aggregations required by decision makers to navigate and to perform analyses in the multidimensional space (using OLAP). Traditionally, these pre-aggregations are modeled by a *lattice* of pre-computed aggregates [3, 8] where:

– each node represents a pre-computed aggregate and
– each edge represents a path for computing aggregates. If the aggregation function used is *distributive* or *algebraic*, the aggregate can be calculated from the directly lower aggregate, while if it is *holistic*, the calculus is from the base relation [8].

To avoid that the lattice is too complex, we simplify the example of the diploma delivery jury. We take into account only two dimensions:

– 'Courses' with two hierarchies 'HCourse_Simple' and 'HCourse_weighted'
– 'Students' with one hierarchy 'HGroup'.

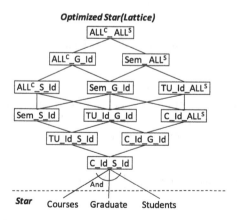

Fig. 5. Classical optimization lattice (We use abbreviations ('Sem' for 'Semester', 'ALL^C' for 'ALL^{DCourses}', 'ALL^S' for 'ALL^{DStudents}'))

Figure 5 shows the lattice of pre-aggregates of the measure 'Rate_Abs'. Each node represents a relation. E.g. the nodes 'TU_Id_S_Id' and 'C_Id_ALL^S' correspond to the following respective relations:

```
TU_Id_S_Id (TU_Id, S_Id, Rate_Abs, Abs_sum, Abs_count)
C_Id_ALL$^S$ (C_Id, Rate_Abs, Abs_sum, Abs_count)
```

In these relations, the attribute 'Rate_Abs' represents the absenteeism rate calculated by the aggregation function AVG. Here, it is a case of algebraic function, so we store intermediate values (the sum 'Abs_sum' and the count 'Abs_count' of occurrences of the absenteeism rate) that will be used to calculate the upper nodes. In the classical approach, contrarily to our proposition, a unique aggregation function is used in the whole lattice for the measure 'Rate_Abs'.

4.3 Extending the Approach with Multiple and Differentiated Aggregations

The flexibility introduced in the conceptual model impacts the lattice.

Increasing the Number of Nodes. In our model, by using the multiple hierarchical and/or differentiated aggregation functions, we can associate the same parameter in different hierarchies with different aggregation functions. This gives different results for the same analysis depending on the used hierarchy. Thus, new nodes compatible with results of all these possible aggregations will be produced in the lattice (Fig. 6). For example, the absenteeism rate 'Rate_Abs' of a TU by groups of students can be calculated by the average function 'AVG(Rate_Abs)' over the hierarchy 'HCourse_Simple' or by the weighted average 'AVG_W(Rate_Abs, Coeff)' on the hierarchy 'HCourse_weighted' (see Fig. 4 (a)); of course each function gives different results.

The number of nodes in the classical lattice (Fig. 5) is calculated by multiplying the number of parameters in each dimension:

$$\text{number of nodes} = \prod_{i=1}^{m} |P^{D_i}|$$

In our model, assuming that each parameter has its own aggregation function, the number of nodes in the lattice (Fig. 6) is calculated by multiplying the sum in each dimension of number of parameters in each hierarchy; here, we must be careful for not count the root parameter of a dimension several times with the different hierarchies:

$$\text{number of nodes} = \prod_{i=1}^{m} \left(\sum_{j=1}^{S_j} (|P^{H_j}| - 1) + 1 \right)$$

Edge Types. The differentiated and multiple aggregation functions involve using different aggregation computations for each edge of the lattice (Fig. 6), contrary to the traditional approach which usually considers only a single aggregation function.

When multiple paths are possible, the less costly path is preferred. The cost function (not detailed here) favors the most effective computation time [15]. However, the use of different aggregation functions on each edge of the lattice makes the cost estimate more complex than in usual lattices.

The possibility of use different aggregation functions for a same measure requires differentiating lattice edges. This typing to indicate the corresponding aggregation function between two nodes. For example, Fig. 4 (a) presents the aggregation schema of the absenteeism rate. Three aggregation functions are used to calculate the absenteeism rate 'Rate_Abs'. For each teaching unit and semester over the hierarchy 'HCourse_weighted', the absenteeism rate takes into account the courses coefficients and teaching units ECTS and uses a weighted function (AVG_W). Thus, in the lattice, it is necessary to distinguish the edges between ('C_Id' and 'TU_Id') and between ('TU_Id' and 'Semester') parameters over the hierarchy 'HCourse_weighted' (that use AVG_W) from the other edges (that use AVG).

In Fig. 6, simple lines correspond to the AVG function and double or triple lines are for AVG_W functions.

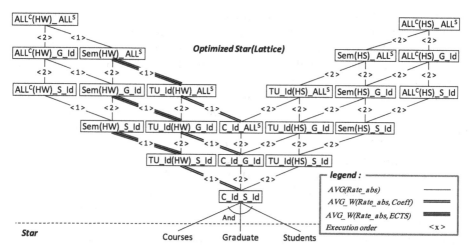

Fig. 6. Lattice with typed edges (Execution orders (< x >) were added to facilitate the understanding of the next impacts (blocking transitivity and pruning the lattice). And we use abbreviations ('HW' for the hierarchy 'HCourse_weighted', 'HS' for the hierarchy 'HCourse_Simple'))

Blocking Transitivity. Constraints (the specific level from which the considered aggregation must be calculated) associated with the aggregation functions have repercussions on the lattice. Edges with a symbol (crosses in a circle in Fig. 7) come from these constraints which require calculating the node from another specific node. It is then forbidden to calculate an upper node using transitivity from lower nodes as it would be in a classical schema. Thus the computing paths are blocked as soon as such an edge is encountered; e.g. the node 'Sem(HS)_S_Id' is calculable from the direct lower node 'TU_Id(HS)_S_Id'; using transitivity, it is also calculable from the lower

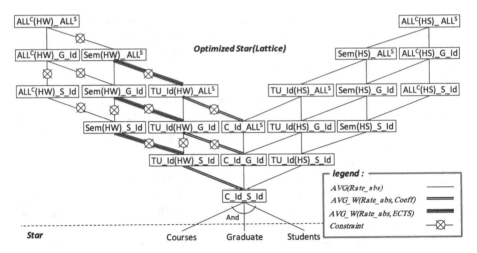

Fig. 7. Lattice with non transitive edges

node 'C_Id_S_Id'. However, the blocked edge resulting from the constraint of the function 'AVG_W(Rate_Abs, ECTS)' which operates on the edge ('TU_Id(HW) _S_Id', 'Sem(HW)_S_Id') blocks the calculation transitivity. Therefore, the node 'Sem (HW)_S_Id' is calculable from the direct lower node 'TU_Id(HW)_S_Id' but not from another lower node (such as 'C_Id_S_Id').

Similarly, the change of execution orders or functions between edges blocks also transitivity. In other words, if all previous edges for a specific edge correspond to different functions or different execution orders, then this edge is non transitive; e.g. the edge ('ALL^C(HW)_S_Id', 'ALL^C(HW)_G_Id') corresponds to the function 'AVG (Rate_abs)' with an execution order of value 2 (see Fig. 6). This edge has a single previous edge ('Sem(HW)_S_Id', 'ALL^C(HW)_S_Id') which corresponds to the same function 'AVG(Rate_abs)' but with an execution order of value 1 (see Fig. 6). Because of the difference between the execution orders, the edge ('ALL^C(HW)_S_Id', 'ALL^C(HW)_G_Id') is not transitive. Therefore, the node 'ALL^C(HW)_ ALL^S' is calculable by transitivity from the node 'ALL^C(HW)_S_Id' but it is not calculable by transitivity from the node 'Sem(HW)_S_Id', because the aggregation schema (Fig. 4 (a)) requires to calculate firstly the absenteeism rates according to the dimension 'Courses' (node 'ALL^C(HW)_S_Id') in order then to calculate absenteeism rate based on the dimension 'Students' ('ALL^C(HW)_ ALL^S').

Figure 7 shows the resulting pre-aggregate lattice. Edges with crossed circle are obtained either from aggregation constraints or from the change of execution orders or aggregation functions between the edges.

Pruning the Lattice. Some paths or edges are invalid; therefore, these can be elimi- nated to reduce the lattice size (Fig. 8). This pruning is possible using the execution order. An edge can be deleted if it is preceded by an edge with a larger execution order (see Fig. 6).

In our example (see Fig. 4 (a)), we cannot apply the weighted average function 'AVG_W(Rate_Abs, Coeff)' on the 'Courses' dimension (with an execution order of value 1) after the function 'AVG(Rate_Abs)' on the dimension 'Students' (with an execution order of value 2) as this would give erroneous results. Thus, to obtain the node 'TU_Id(HW)_G_Id' (absenteeism rate by group and TU on the hierarchy 'HCourse_weighted'), it is impossible to calculate it from the node 'C_Id_G_Id' (absenteeism rate by group and course on the hierarchy 'HCourse_weighted'). Therefore, the edge between 'C_Id_G_Id' and 'TU_Id(HW)_G_Id' can be deleted.

Figure 8 shows the final controlled pre-aggregate lattice after deleting the invalid edges.

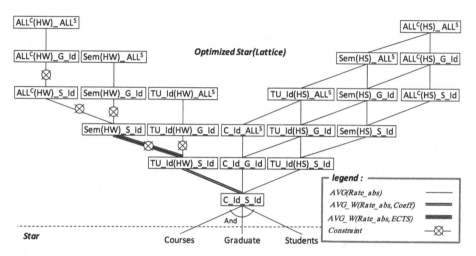

Fig. 8. Controlled pre-aggregate lattice (with pruned edges)

Modifying Edges. In our model, we have proposed a mechanism of aggregation constraint to fix the valid aggregation level from which a higher level is calculated. This valid level is not necessarily the one directly lower level. We express this case when we use a constraint value other than 0 (the aggregation can be calculated from any lower level) or - 1 (the aggregation can only be calculated from the level directly below the selected one). Such constraints imply possible path changes in the lattice.

In our example, the absenteeism rate by semester on the hierarchy 'HCourse_ weighted' is calculated from the absenteeism rates by TU (constraint value = -1) (see Fig. 4 (a)). In case we had chosen to calculate this rate by semester from the rates by courses, the constraint would have been assigned to -2 and the lattice would have been as Fig. 9.

5   Validations

To demonstrate the feasibility of our approach, we have produced a prototype: *OLAP-Multi-Function*, described hereafter. We validate our proposal by overcoming the limits suffered by the software "Business Objects" with our prototype. Finally, experiments based on our prototype are detailed.

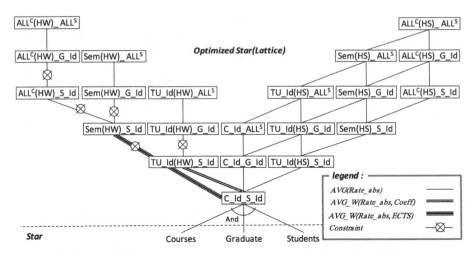

Fig. 9. Lattice with constraint = -2

5.1 OLAP-Multi-Function Prototype

Our prototype was implemented using Java 7 on top of the Oracle 12g DBMS. It allows designing a MDB with differentiated and multiple aggregation functions as well as supervising the OLAP manipulations carried out by analysts using a graphical representation.

Prototype Architecture. The main functionality of *OLAP-Multi-Function* (Fig. 10) is visualizing and facilitating the integration of aggregation functions in the multidimensional model. It is based on a set of graphic interfaces (Constructor) for defining the four types of aggregation functions (general, multiple dimensional, multiple hierarchical, differentiated), their execution orders and aggregation constraints.

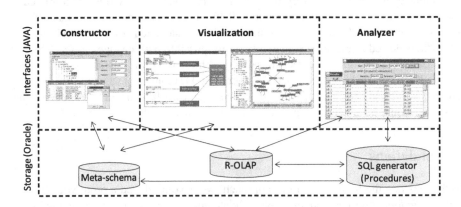

Fig. 10. Prototype architecture

The structural schema is displayed as a constellation graph based on graphic formalisms of facts, dimensions, and hierarchies introduced in [21, 22]. Different aggregation schemas are visualized in the form of a hyperbolic graph. For querying, the analyst selects the measure and the desired aggregation levels. After validation, *OLAP-Multi-Function* automatically calculates the result and presents it in the form of an R-OLAP table.

The storage level includes two databases. The first one contains the meta-schema that describes the structural elements of the multidimensional schema (facts, dimensions and hierarchies) as well as the aggregation functions, execution orders and aggregation constraints to build valid and coherent SQL queries (for more information about the meta-schema, see [12]). The second one contains facts and dimensions data implemented with the R-OLAP model.

SQL Queries Generator. To supervise the analysis, the prototype has a SQL query generator. The analyst configures the calculations to be done: the user must specify the measure and the desired aggregation levels. The generator translates interactions into SQL scripts executable in the context of an R-OLAP implementation. The generation process consists of the four following steps, described using a BPMN diagram (Business Process Modeling Notation) in Fig. 11:

1. Detecting tables of the logical R-OLAP model: this step identifies the tables used to store analysis data.
2. Determining aggregation functions: using the meta-schema and the required aggregation levels, this step identifies aggregation functions to be applied to perform the analysis.
3. Simplifying aggregation functions: this step is for detecting possible redundant calculations, i.e. a needless repetition of an aggregation function
4. Generating the SQL script: from the meta-schema and the previous steps, this step generates the final SQL query. It sends it to the DBMS that calculates the query and returns the results to the prototype.

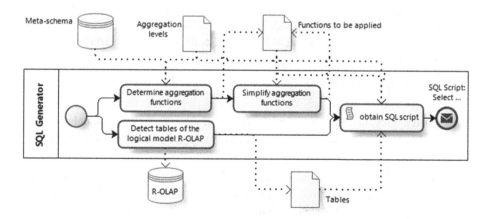

Fig. 11. SQL queries generator (shown in BPMN)

5.2   Discussing

We present in this section the advantages of our prototype *OLAP-Multi-Function* over one of the most used commercial tool: "Business Objects" (BO).

Business Objects. According to our knowledge, the major limit of BO is to use only one aggregation function for each measure. To know how far we can overcome this problem, we have applied our example (Fig. 2) in BO. We associated the measure 'Avg_Mark' with the aggregation function AVG. Thus, we can perform all possible analyses on dimensions 'Students' and 'Dates'. For example, we can analyze the average marks of courses ('C_Id' level) by periods of five years ('period-5' level and 'ALL' level on the 'Students' dimension). This analysis can be performed by the following SQL query:

```
SELECT C.C_Id, D.Period-5, AVG(G.Avg_Mark) AS Avg_Mark
FROM COURSES C, DATES D, GRADUATE G
WHERE G.C_Id = C.C_Id AND G.Academic_year = D.Academic_year
GROUP BY C.C_Id, D.Period-5
```

But for analyzing the data along the dimension 'Courses', we use a non-standard aggregation function: 'AVG_W' (weighted average). To solve this problem, there are two proposals:

- The use of a calculated measure: this proposal means defining a new measure (AVG_Mark_TU) calculated by Eq. 1, defined in Sect. 1.2 (Fig. 12). The problem with this proposal is that this equation ("Select:" in Fig. 12) will not be used to calculate the measure at the TU level but at the base level ('C_ID'), then to calculate the measure at the TU level, its own aggregation function will be used to aggregate the values.
- The use of a variable: the advantage of this proposal is that the variable can use values of an aggregated measure contrary to the calculated measure that use only the base values. The problem is that if the variable uses values other than the measure, these values must be used in the analysis, otherwise there will be errors; e.g. the variable 'AVG_Var' (Fig. 13) is calculated by Eq. 1 where the values of 'Coeff' are not used in the analysis, hence the errors. To overcome this problem and get the requested results, we can define two new measures: the first M1 = SUM (Coeff * AVG_Mark); the second M2 = SUM(Coeff) and then the variable becomes AVG_Var = M1 /M2.

Thus, by using variables, we can calculate:

1. A non-standard function,
2. A second aggregation function from the results of the main function associated with the measure. This is similar to associating two aggregation functions with one measure.

The limits of the use of a variable are when that variable is used for a specific level (as the variable 'AVG_Var' in our example); then there is no constraint that forbids the user to use it for a different level and that would give a wrong result. Another limit is that we cannot use a variable to calculate another variable otherwise there will be errors

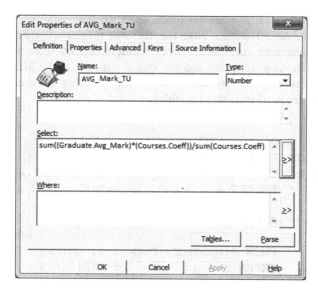

Fig. 12. Use of a calculated measure in a BO query

	Nom	ST1		ST2	
Semestre	TUTitle	AVG_Mark	AVG_Var	AVG_Mark	AVG_Var
5	TU1	12,00	#IERR	12,67	#IERR
5	TU2	11,00	#IERR	13,00	#IERR
6	TU3	15,00	#IERR	12,00	#IERR
6	TU4	10,50	#IERR	10,00	#IERR

Fig. 13. Use of a variable in a BO report

(Fig. 13). Thus we cannot use the variable 'AVG_Var' to calculate the average marks at the Semester level.

OLAP-Multi-Function. Our prototype integrates several aggregation functions for the same measure in the multidimensional model. It overcomes the two principal limits of BO: the use of non-standard functions and the use of several aggregation functions. To use the non-standard weighted average function 'AVG_W', a generic aggregation function was implemented:

- An Oracle object type (class) was used to implement the four routines of the Data cartridge interface ODCIAggregate: ODCIAggregateInitialize, ODCIAggregateIterate, ODCIAggregateMerge and ODCIAggregateTerminate. These methods correspond to internal operations that each aggregation function performs (respectively Initialize, Iterate, Merge and Terminate).
- Then, our aggregation function 'AVG_W' was created to compute a weighted average based on our previous object type. This function takes one parameter composed of the data to aggregate and the weight (TYPE data_weighted AS OBJECT (value NUMBER, weight NUMBER)).

In order to use several aggregation functions in the same analysis, our SQL generator can generate nested queries. Note that the SQL queries are generated using an interface where the user manipulates only multidimensional concepts. Thus, the complexity of both the aggregation and the logical structure of the MDB are hidden. E.g. the SQL query generated by our prototype for analyzing the average marks by semester and by group of students is as follows:

```
SELECT G_Id, Semester, AVG(Avg_Mark) AS Avg_Mark
FROM ( SELECT G_Id, Semester, S_Id, Academic_year,
       AVG_W(DATA_WEIGHTED(Avg_Mark, ECTS)) AS Avg_Mark
       FROM ( SELECT C.C_Id, C.Semester, S.S_Id,
              D.Academic_year,C.TU_Id, C.ECTS,
              AVG_W(DATA_WEIGHTED(G.Avg_Mark, C.Coeff))
                                            AS Avg_Mark
              FROM STUDENTS S, COURSES C, DATES D, GRADUATE G
              WHERE G.S_Id = S.S_Id AND G.C_Id = C.C_Id
              AND G.Academic_year = D.Academic_year
              GROUP BY S.G_Id, C.Semester, S.S_Id,
                       D.Academic_year, C.TU_Id, C.ECTS)
       GROUP BY G_Id, Semester, S_Id, Academic_year)
GROUP BY G_Id, Semester
```

5.3   Experiments

The SQL query generator serves as an experimental platform for which we show a series of experiments.

Experiment 1. The first experiment is intended to study the impact of our proposal on the execution time of OLAP analysis queries.

Collection: to our knowledge, there are no benchmarks that use for a same measure, different aggregation functions according to analysis axes, hierarchies and aggregation levels. Therefore, we use data related to the diploma delivery jury; the data grouping size on the dimension 'Courses' is five, i.e. each instance of a higher level corresponds to five instances of lower level (for example, each semester has five TUs).

Protocol: we observe the execution time (in seconds) in accordance with the number of tuples of the fact (from two to ten millions) of three queries:

- The first query aggregates average marks at the TU level. It uses (as in the classical model) a single aggregation function ('AVG_W'),
- The second query aggregates average marks at the semester level. It is based on two aggregation functions ('AVG_W' twice),
- The third query aggregates average marks at the 'ALL' level. It requires three aggregation functions ('AVG_W' twice and 'AVG' once).

We chose these three queries to present the impact of using several functions (second and third queries) compared with the classical model that uses a single function (first query).

Results: Fig. 14 (right) shows the execution time of the three queries. Queries execution times increase regularly with the number of tuples. The distance between the curves of the first query (aggregation in the classical model) and the second query (aggregation in our proposed model) represents the overhead time of our model required to apply the second aggregation function. This time is approximately 5 % of the total query execution time. The additional time to apply the third function seems to be non-remarkable (the curve of third query is nearly on top of the curve of second query). In fact, this phenomenon is related to the data volume that decreases with the functions previously applied. Thus, when calculating the third function, the data volume is significantly reduced compared with the initial volume.

Experiment 2. The second experiment aims at studying the relationship between the execution time and the data grouping size. By grouping size, we mean the number of values of a lower parameter that are grouped into one value of a higher parameter.

Collection: we work on two different versions of our example of the diploma delivery jury; the first one with data grouping size 2 on the dimension 'Courses' and the second one with data grouping size 5.

Protocol: we observe the execution time in accordance with the number of tuples and the size of the data grouping of the four queries:

- Two queries at the TU level (one with a grouping size of 2 and the other with5) that use an aggregation function.
- Two queries at 'ALL' level (one with a grouping size of 2 and the other with 5) that use three aggregation functions.

Results: Fig. 14 (left) shows the execution time of the four queries. The execution time of queries with grouping size 5 is less than that of the queries with a grouping size of 2. We note that the query execution time seems mainly influenced by the grouping size. Thus, the query with grouping size 2 and a single aggregation function (TU (2)) is more

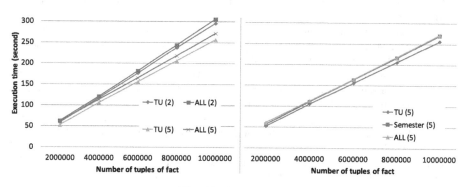

Fig. 14. Experiments

expensive in terms of computing time than the query with a grouping size of 5 despite three aggregation functions (ALL (5)). The grouping size appears to have a crucial impact on the execution time.

6 Conclusion and Future Work

This paper defines a conceptual multidimensional data model flexible enough to allow the designer to specify differentiated and multiple aggregations. *Multiple*, as the same measure can be aggregated by several aggregation functions according to analysis axes or hierarchies and *differentiated* as these aggregations may vary, depending on the aggregation level. Furthermore, the model is expressive enough to check the function calculations validity. Aggregation constraints define the level from which the aggregation should be calculated. The execution order defines the necessary order between non-commutative aggregation functions.

This model is based on a two-level graphical formalism: the structural schema describes the multidimensional structures while hiding the aggregation complexity and aggregation schemas detail the aggregation mechanisms for each measure.

At the logical level, the implementation can be optimized by a controlled lattice of pre-computed aggregates, where invalid edges can be pruned.

We plan to continue our work by revisiting algorithms that compute pre-aggregates, adapting them to our model and studying the effects of changes in the lattice when selecting nodes for improving performance. We also plan to study OLAP manipulation operators on our model.

References

1. Abelló, A., Samos, J., Saltor, F.: YAM2: a multidimensional conceptual model extending UML. Inf. Syst. **31**(6), 541–567 (2006)
2. Boulil, K., Bimonte, S., Pinet, F.: Un modèle UML et des contraintes OCL pour les entrepôts de données spatiales. De la représentation conceptuelle à l'implémentation. In: Ingénierie des Systèmes d'Information (ISI), vol. 16(6), pp. 11–39 (2011). (In French)
3. Chaudhuri, S., Dayal, U.: An overview of data warehousing and OLAP technology. SIGMOD Rec. **26**(1), 65–74 (1997)
4. Cuzzocrea, A.: Providing probabilistically-bounded approximate answers to non-holistic aggregate range queries in OLAP. In: 8th International Workshop on Data Warehousing and OLAP (DOLAP 2005), pp. 97–106 (2005)
5. Cuzzocrea, A., Furfaro, F., Masciari, E., Saccà, D., Sirangelo, C.: Approximate query answering on sensor network data streams. In: Stefanidis, A., Nittel, S. (eds.) GeoSensor Networks, pp. 53–72. CRC Press, Boca Raton (2004)
6. Cuzzocrea, A., Saccà, D.: Balancing accuracy and privacy of OLAP aggregations on data cubes. In: 13th International Workshop on Data Warehousing and OLAP (DOLAP 2010), pp. 93–98 (2010)
7. Golfarelli, M., Maio, D., Rizzi, S.: Conceptual design of data warehouses from E/R schemes. In: International Conference on HICSS 1998, vol. 7, pp. 334–343 (1998)

8. Gray, J., Bosworth, A., Layman, A., Pirahesh, H.: Data cube: a relational aggregation operator generalizing group-by, cross-tab, and sub-total. In: International Conference on ICDE 1996, pp. 152–159 (1996)
9. Gyssens, M., Lakshmanan, L. V. S.: A foundation for multi-dimensional databases. In: International Conference on VLDB 1997, pp. 106–115 (1997)
10. Harinath, S., Zare, R., Meenakshisundaram, S., Carroll, M., Guang-Yeu Lee, D.: Professional Microsoft SQL Server Analysis Services 2008 with MDX. Wiley Publishing, Indianapolis (2009)
11. Hassan, A., Ravat, F., Teste, O., Tournier, R., Zurfluh, G.: Differentiated multiple aggregations in multidimensional databases. In: Cuzzocrea, A., Dayal, U. (eds.) DaWaK 2012. LNCS, vol. 7448, pp. 93–104. Springer, Heidelberg (2012)
12. Hassan, A., Ravat, F., Teste, O., Tournier, R., Zurfluh, G.: Agrégations multiples différentiées dans les bases de données multidimensionnelles. In: Ingénierie des Systèmes d'Information (ISI), vol. 18(2), pp. 75–102 (2013). (In French)
13. Kimball, R., Ross, M.: The Data Warehouse Toolkit: the Definitive Guide to Dimensional Modeling, 3rd edn. John Wiley & Sons, NY (2013). ISBN 978-1-118-53080-1
14. Jaecksch, B., Lehner, W.: The planning OLAP model - a multidimensional model with planning support. In: Cuzzocrea, A., Dayal, U. (eds.) DaWaK 2011. LNCS, vol. 6862, pp. 14–25. Springer, Heidelberg (2011)
15. Kotidis, Y., Roussopoulos, N.: DynaMat: a dynamic view management system for data warehouses. In: International Conference on SIGMOD 1999, pp. 371–382 (1999)
16. Lujàn-Mora, S., Trujillo, J., Song, I.Y.: A UML profile for multidimensional modeling in data warehouses. Data Knowl. Eng. **59**, 725–769 (2006)
17. Mazón, J.N., Lechtenbörger, J., Trujillo, J.: A survey on summarizability issues in multidimensional modelling. Data Knowl. Eng. **68**, 1452–1469 (2009)
18. Oliveira, R., Rodrigues, F., Martins, P., Moura, J.P.: Extending the dimensional templates approach to integrate complex multidimensional design concepts. In: Cuzzocrea, A., Dayal, U. (eds.) DaWaK 2011. LNCS, vol. 6862, pp. 26–38. Springer, Heidelberg (2011)
19. Pedersen, T.B., Jensen, C., Dyreson, C.: A foundation for capturing and querying complex multidimensional data. Inf. Syst. **26**, 383–423 (2001)
20. Prat, N., Wattiau, I., Akoka, J.: Representation of aggregation knowledge in OLAP systems. In: the 18th European Conference on Information Systems ECIS. (2010)
21. Ravat, F., Teste, O., Tournier, R., Zurfluh, G.: Graphical querying of multidimensional databases. In: Ioannidis, Y., Novikov, B., Rachev, B. (eds.) ADBIS 2007. LNCS, vol. 4690, pp. 298–313. Springer, Heidelberg (2007)
22. Ravat, F., Teste, O., Tournier, R., Zurfluh, G.: Algebraic and graphic languages for OLAP manipulations. Int. J. Data Warehous. Min. **4**(1), 17–46 (2008)
23. Silva, J., Times, V.C., Salgado, A.C.: A set of aggregation functions for spatial measures. In: 11th International Workshop on Data Warehousing and OLAP (DOLAP 2008), ACM. ISBN: 978-1-60558-250-4, pp. 25–32 (2008)
24. Torlone, R.: Conceptual multidimensional models. In: Rafanelli, M. (ed.) Multidimensional Databases: Problems and Solutions, pp. 69–90. IGI Publishing Group, PA (2003)
25. Vassiliadis, P., Sellis, T.K.: A survey of logical models for OLAP databases. SIGMOD Rec. **28**(4), 64–69 (1999)
26. Vassiliadis, P., Simitsis, A., Skiadopoulos, S.: Modeling ETL activities as graphs. In: International Conference on DMDW 2002, pp. 52–61 (2002)
27. Vassiliadis, P., Skiadopoulos, S.: Modelling and optimisation issues for multidimensional databases. In: Wangler, B., Bergman, L.D. (eds.) CAiSE 2000. LNCS, vol. 1789, pp. 482–497. Springer, Heidelberg (2000)

MIRABEL DW: Managing Complex Energy Data in a Smart Grid

Laurynas Šikšnys$^{(\boxtimes)}$, Christian Thomsen, and Torben Bach Pedersen

Department of Computer Science, Aalborg University,
Aalborg, Denmark
{siksnys,chr,tbp}@cs.aau.dk

Abstract. In the MIRABEL project, a data management system for a *smart grid* is developed to enable smarter scheduling of energy consumption such that, e.g., charging of car batteries is done during night when there is an overcapacity of *green energy* from windmills etc. Energy can then be requested by means of *flex-offers* which define flexibility with respect to time, amount, and/or price. In this paper, we describe MIRABEL DW, a data warehouse (DW) for the management of the large amounts of complex energy data in MIRABEL. We present a unified schema that can manage data both at the level of the entire electricity network and the level of individual nodes, such as a single consumer node. The schema has a number of complexities compared to typical DW schemas. These include *facts about facts* and *composed non-atomic facts* and unified handling of different kinds of flex-offers and time series. We also discuss alternative data modeling strategies and how specialized variants of the generic schema can be used by different node types while we maintain compatibility and consistency between them. Finally, we present typical queries from the energy domain and a performance study.

1 Introduction

More and more *green energy* is being produced by renewable energy sources (RES) such as windmills. It is, however, not possible to store larger amounts of energy and use it later. Therefore, there often is an unused capacity, e.g., during nights when most consumers sleep, but not enough green energy during day hours when most consumers are active. The EU FP7 project MIRABEL (Micro-Request-Based Aggregation, Forecasting, Scheduling of Energy Demand. Supply and Distribution) [14] addresses this challenge by proposing a "data-driven" solution for balancing supply and demand utilizing their flexibilities. Flexible demand such as for dishwashers and charging an electric vehicle can often be shifted to a time when green energy is available. Non-flexible demand such as lights, TV, or cooking stoves must still be satisfied at demand-time. In the MIRABEL-settings, a consumer offers a so-called *flex-offer* [2,16] for every intent of flexible energy demand. The flex-offer must describe when and how much energy is needed and how flexible the demand is in time and amount. Likewise, a producer can offer a flex-offer for every intent of energy supply.

© Springer-Verlag Berlin Heidelberg 2015
A. Hameurlain et al. (Eds.): TLDKS XXI, LNCS 9260, pp. 48–72, 2015.
DOI: 10.1007/978-3-662-47804-2_3

The different flex-offers can then be accepted (or rejected if they cannot be fulfilled) and scheduled for execution at a given time. There will be extremely large quantities of such flex-offers and they cannot be scheduled individually. Instead flex-offers are *aggregated* into larger flex-offers which become scheduled and then *disaggregated* into the smaller flex-offers again [16]. To enable this, there will be smart *nodes* at both consumer sites and producer sites in the electricity grid which we denote a *smart grid*.

There is a strong need for efficient data management in these nodes. In this paper, we present *MIRABEL DW* which is a data warehouse (DW) for the management of large amounts of complex energy data in the MIRABEL project. This paper is the first to present a DW schema for the important domain of energy data. The schema can represent different "actors" in different "roles" as defined by the "Harmonised Electricity Market Role Model" [6] as well as (individual and aggregated) flex-offers, and time series. In the future, the managed data is to be distributed over millions of nodes [2] in non-traditional ways. In the paper, we focus on a DW on a single node, but present a unified schema that can manage data both at the level of the entire electricity network and the level of individual nodes, such as a single consumer node. Compared to typical DW schemas, the schema has a number of complexities which we discuss in the paper. These include *facts about facts* and *composed non-atomic facts* and unified handling of different kinds of flex-offers and time series. We also discuss alternative data modeling strategies that use denormalization and arrays, respectively. We also discuss so-called *specializations* which allow certain variants of the generic unified schema to simplify data management in different node types which, e.g., can have limited hardware resources. Further, we present typical queries from the energy domain and a performance study that compares the described schemas with the denormalized and array-based alternatives, and the specialized schemas.

The rest of the paper is organized as follows: Our representations of flex-offers, time series and actors are presented in Sects. 2, 3, and 4, respectively. These parts together form the full schema which is presented in Sect. 5. Section 6 presents specializations of the generic schema to simplify data management at different node types. Examples of analytical queries on the schema are given in Sect. 7. A performance study is given in Sect. 8. Previous work related to this is presented in Sect. 9 before the concluding remarks and pointers to future work which are given in Sect. 10.

2   Modeling of Flex-Offers

In this and the following two sections, we first present the data model we use in MIRABEL DW. Then we discuss the non-standard and advanced techniques that are applied in the modeling.

2.1   Data Model

To represent MIRABEL's flex-offers (both aggregated and non-aggregated) is an essential task for MIRABEL DW. This is done by means of the tables shown

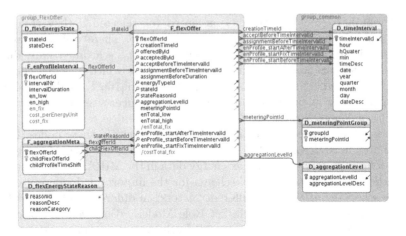

Fig. 1. Tables for representing flex-offers

in Fig. 1. We first describe the dimensions (which are recognized by the prefix D_ in their table names) and then the fact tables (recognized by the prefix F_ in their names). All dimension tables have surrogate keys with names ending with Id. The possible states for a flex-offer (such as "offered", "accepted", and "rejected") are represented in the dimension D_flexEnergyState. A flex-offer has its state for a certain reason (for example, a flex-offer becomes rejected if the offered price is too high). The possible reasons are represented in the dimension D_flexEnergyStateReason. As we expect few generic reason categories (e.g., "Price too high") and many more specific reason descriptions (e.g. "Price (499.50 euros) too high") to exist, we have columns for both the generic categories and the specific reasons such that a hierarchy exists. In MIRABEL DW, we represent time by discretized time intervals. This is done by D_timeInterval which represents 15 min intervals (for now; other interval lengths can be chosen if needed). Flex-offers are always related to at least one metering point (at the location where the energy is to be consumed or produced), but if a flex-offer is aggregated, it will be associated with many metering points. To capture this, D_meteringPointGroup is used as bridge table [9] between the fact table and D_meteringPoint which represents the individual metering points. To represent the aggregation level of a flex-offer, D_aggregationLevel is used.

The fact table F_flexOffer holds flex-offer facts. It references all the previously described dimension tables. There are six foreign keys to D_timeInterval to represent different times such as when the flex-offer was created and when it at the latest has to be assigned etc. These foreign keys thus all represent an absolute time. There is also an attribute assignmentBeforeDuration which holds a time span telling how long before the actual execution time the assignment must take place.

Further, F_flexOffer references D_legalEntityRole (explained later) twice to represent who offered and accepted the flex-offer, respectively. Only the current information about a flex-offer is held; if a flex-offer is modified, the old fact is overwritten. There are measures to hold the lowest and highest amount of energy required by the flex-offer as well as a measure to hold the "fixed" amount of energy that becomes accepted. Further, a measure holds the total cost of the fix. Finally, each represented flex-offer is given a unique identifier in the attribute flexOfferId which technically is a degenerate dimension.

Information about the profile intervals of flex-offers is represented in the fact table F_enProfileInterval. This fact table only has a single foreign key which references the unique flexOfferId in F_flexOffer. The imported value together with a sequential intervalNr forms the primary key for F_enProfileInterval. The reason for this design is that a single flex-offer can have many profile intervals. For each represented profile interval, there is a duration specifying how many time units the profile interval spans over, and both the lowest and highest amount of energy needed in this interval. When the flex-offer becomes fixed, the actual amount of energy in the interval and the price for this energy also becomes represented. An alternative to this design would be to represent the measures of F_enProfileInterval in *arrays* in F_flexOffer such that all data about a given flex-offer would be represented in a single fact. Yet another alternative would be to represent all attributes of F_enProfileInterval in F_flexOffer, i.e., denormalize the data and have one (wide) fact in F_flexOffer for each profile interval. (For space reasons, we do not show the alternative schemas in figures).

As flex-offers can be aggregated into larger flex-offers, we also introduce the table F_aggregationMeta which references F_flexOffer twice to point to the aggregating "parent flex-offer" and the smaller "child flex-offer" which has been aggregated, respectively. Profiles of each child flex-offer can be shifted relatively to the profile start of the parent flex-offer when aggregating child flex-offers into the parent. Therefore, for every child flex-offer, the childProfileTimeShift attribute indicates the amount of time units the profiles of the child flex-offer has been shifted in the aggregated flex-offer. This information is used in the disaggregation.

2.2   Modeling Challenges

The fact table F_flexOffer is the central fact table for representation of flex-offers. It is, however, also used as a dimension table in the sense that each fact has a unique ID such that F_enProfileInterval and F_aggregationMeta can reference F_flexOffer and in effect store *facts about facts*. Considering F_flexOffer and F_enProfileInterval, it can even be discussed *what* a fact is. An energy profile interval (in this context) always belongs to a flex-offer and any meaningful flex-offer has an energy profile interval (a flex-offer for zero consumption/production at an undefined point in time is hardly interesting). It could be argued that a single fact is represented by a single row in F_flexOffer and many rows in F_enProfileInterval. Unlike traditional DW schemas, we thus have non-atomic *composed* facts. As pointed out above, we could alternatively have modeled this by using arrays in F_flexOffer to hold the measures that currently are represented in F_enProfileInterval. This would, however, make it more cumbersome to compare different measures (e.g., en_low with the minimum energy requirement to en_fix with the assigned energy) as the interval position currently represented by intervalNr only would be implicitly represented by the position in the array. The denormalized variant (with a fact in F_flexOffer for each profile interval) would increase redundancy dramatically.

Another interesting aspect of MIRABEL DW is how it represents facts for both non-aggregated and aggregated flex-offers in a unified way. The aggregation is unlike traditional aggregation since the parent flex-offer contains other flex-offers that can be shifted within the parent flex-offer. We call the contained flex-offers *shiftable child facts*.

3 Modeling of Time Series

3.1 Data Model

In MIRABEL DW, time series are represented by means of the tables shown in Fig. 2. It is necessary to be able to represent time series of various types, for now energy, power, and price. To represent these general classes, we use the D_typeClass dimension table. Apart from its surrogate key, it has the attribute typeClassDesc which holds a textual description of the time series type (such as "Energy") and the attribute unit which holds the unit of measurements (such as "kWh"). Instances of the general types are represented in the table D_type. For example, an instance of the "Energy" class is "Energy-Metered-Production-RES-Wind". D_type references D_typeClass to represent the hierarchy between

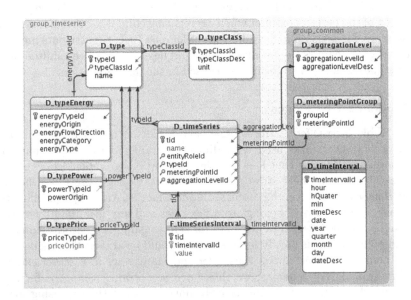

Fig. 2. Tables for representing time series

types and type classes. For different types of time series, it is, however, necessary to store different information. Therefore, we introduce the tables D_typeEnergy, D_typePower, and D_typePrice to hold the attributes that are relevant for the different types. These tables supplement, but cannot replace, D_type. The reason is that we need a single table to reference from D_timeSeries to represent the type of the time series in question. Thus D_type is referenced from D_timeSeries, but the special attributes for an energy time series are represented in D_typeEnergy. The latter table has columns to describe the origin of the time series (e.g. "Metered" or "Forecasted"), the flow direction (i.e., if it is production or consumption), the category (e.g., energy from renewable energy sources), and the type of energy (e.g. "Wind"). The design is likely to evolve in the future. For example, there is a traditional hierarchy where types roll up into categories that roll up into flow directions. A more advanced hierarchy is, however, needed to represent hybrid energy types like "At least 90 % energy from renewable energy sources and the rest produced from coal".

D_timeSeries holds a single entry for an entire time series. For each represented time series, there is a unique ID tid and a name may be given. Further, D_timeSeries references D_type (as previously described), D_aggregationLevel to represent the level of aggregation of the time series, and D_meteringPointGroup to represent which meters the time series describes. Thus, D_timeSeries is mainly used to relate different dimension values that describe the represented time series. The values of the time series are, however, represented in the fact table F_timeSeriesInterval. This table references D_timeSeries to identify the time series a value belongs to and D_timeInterval to identify the time instant when the value occured. Finally, the table holds the value itself as the measure. A fact thus exists for each value in each time series. It can, however, also be argued that a fact consists of what it represented in F_timeSeriesInterval *and* what is represented in D_timeSeries which – apart from a possible name – only points out to other dimensions.

3.2   Modeling Challenges

Similarly to the representation of flex-offers, our representation of time series also leads to compound facts where one fact can be considered to be made up of parts in different tables (D_timeSeries and F_timeSeriesInterval). Actually, an alternative design is to merge F_timeSeriesInterval into D_timeSeries such that the values instead are represented in an array, meaning that a single time interval (and all its values) only would result in one fact. Yet another alternative is to merge D_timeSeries and F_timeSeriesInterval and have a row for each value in a time series. There are thus different possible ways to represent the complex sequence-facts arising from time series. We choose the model in Fig. 2 since it both reduces complexity (compared to the first alternative where two arrays must be processed to find the value for a given time instant) and redundancy (compared to the second alternative where there is very wide fact for each value in the time series).

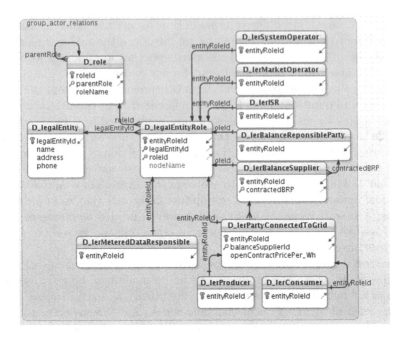

Fig. 3. Tables for representing different actors/roles

In our modeling of time series, the schema is neither a traditional star schema nor a snowflake schema. One reason for this is of course the compound facts discussed above. Another reason is the support for different types of time series for which different attributes are needed. We have different tables that reference D_type which also is the dimension table referenced from the fact table. Consider for example D_typeEnergy which represents attributes that are relevant for energy time series. An alternative design would be to join all these D_type* tables into one dimension table, but for every dimension member many attribute values would then be NULL.

4   Modeling of Different Actors and Market Areas

4.1   Data Model

Many different entities are involved in different roles in energy trading and network operation. We represent the needed actors from the "Harmonised Electricity Market Role Model" [6] by means of the tables in Fig. 3

The table D_role represents roles such as "Producer" and "Consumer". A role can belong to another parent role and this is captured by a self-reference. For example, the parent role of both "Producer" and "Consumer" is "Party Connected To Grid". Legal entities are represented by D_legalEntity. To capture when a certain legal entity plays a certain role (a single legal entity can play several roles), we use D_legalEntityRole. This table references both D_role and D_legalEntity. Further, it has an attribute to hold a unique ID for a given legal entity playing a given role. We include this ID as it makes it easy to point to a legal entity in a certain role. We do exactly that from a number of tables as shown in Fig. 3. For each role, there is a specialized table that (directly or indirectly through another table) references D_legalEntityRole. Some of them, like D_lerSystemOperator, are simple and do only have one attribute which is a reference to this ID. The specialized table can be referenced and it is then explicit what kind of role is referenced. For example, the table D_lerSystemOperator is referenced from D_marketBalanceArea as shown in Fig. 5. A slightly more complex example is D_lerPartyConnectedToGrid which references D_legalEntityRole and also D_lerBalanceSupplier to represent that a party connected to the grid always is so through a balance supplier. Further, D_lerPartyConnectedToGrid is itself referenced from its specializations, D_lerProducer and D_lerConsumer.

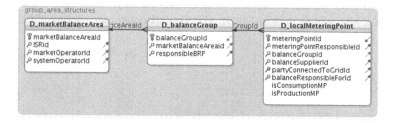

Fig. 4. Tables for representing market areas

Finally, we have tables to represent market areas as shown in Fig. 4. D_localMeteringPoint represents the meters that are connected to the grid. Such meters are installed both at the producer and consumer sites. D_localMeteringPoint references four different specializations of D_legalEntityRole. Further, it references D_balanceGroup which in turn references D_marketBalanceArea which hierarchically groups metering points.

4.2   Modeling Challenges

To the best of our knowledge, this is the first paper to describe a DW for the complex concepts of actors and roles in the "Harmonised Electricity Market Role Model" [6]. Our model captures both how legal entities can play different roles and how roles can be parts of other roles. This is captured by the tables D_legalEntity, D_role, and D_legalEntityRole. In addition to these tables, a (narrow) table has been added for each role a legal entity can play (see the D_ler* tables). It is then possible to represent attributes that are only relevant for certain roles such as done for D_lerBalanceSupplier. Further, when foreign keys reference these tables (instead of just referencing D_legalEntityRole), it is explicit what kind of role playing is referenced and it helps to avoid mistakes where, e.g., a balance supplier is referenced where a balance responsible party actually should have been referenced. We note that if no special attributes must be stored for the different roles, then instead of storing the D_ler*'s as physical tables, they can be views selecting from D_legalEntityRole. This reduces the risk of mistakes further and makes maintenance of them automatic.

5   The Full Schema

To summarize the previous descriptions, the full schema for MIRABEL DW is shown in Fig. 5. The schema can capture the (needed) roles from the Harmonised Model [6] as well as the "actor configurations" where different actors play different roles. The schema also includes specializations of legal entities. Further, the schema can capture different kinds of time series as complex sequence facts. The schema is thus general enough to hold all the data that is needed in the MIRABEL project. It should, however, be noted that no single node is intended to hold all data. Instead, a node should only hold data that is relevant for the site where it is installed. For an end-consumer this would typically

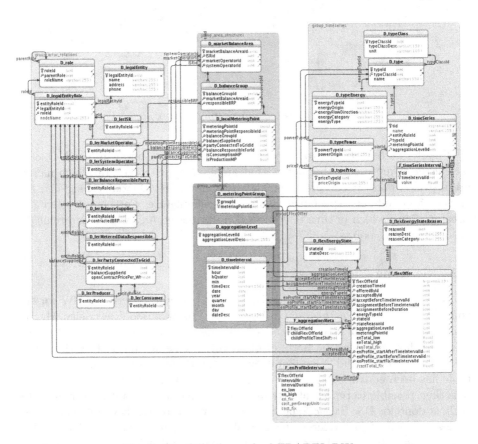

Fig. 5. The full schema for MIRABEL DW

be her own non-aggregated flex-offers and time series about metered energy. For a balance responsible party buying electricity on the market and selling it to end-consumers, it would include both aggregated and non-aggregated flex-offers, forecasted and metered time series, and market areas. The data will thus be distributed accordingly to the roles played by the owners of the nodes. The data will also be at different aggregation levels such that some nodes have detailed data while others have more aggregated data. A consumer will know the details of her flex-offers, i.e., when she has requested energy and how much. For a balance responsible party, the individual non-aggregated flex-offers and end-users generating may not be known, but the aggregated information will be known, e.g., that x MWhs must be produced in a given time interval. Note that the different nodes can use the same schema. The distribution of data is illustrated in Fig. 6 which shows different kinds of nodes. Non-aggregated flex-offers are shown as small, shaded boxes. Note that the different nodes do not represent the same flex-offers. A single node only represents the flex-offers that are relevant to its owner. Aggregated flex-offers are shown as larger, filled boxes in Fig. 6. Note also that although the nodes distribute the data and some represent non-aggregated flex-offers and others only represent aggregated flex-offers, they can use the same schema. As described in the following section, another possibility is to allow the different kinds of nodes to use specialized schemas.

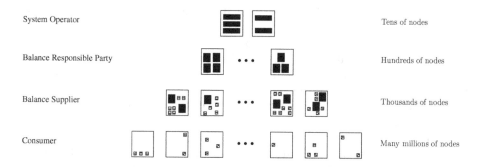

Fig. 6. Data distribution in MIRABEL DW

6 Schema Specializations

The schema in Fig. 5 is generic and can be used in all kinds of nodes in MIRABEL. It is, however, not all kinds of nodes that need to store all kinds of data. Consider, for example, a node installed at an end-consumer's site (i.e., the lowest level in the hierarchy of nodes) on limited hardware resources. Such a node does not store aggregated flex-offers; it only knows the consumer's own flex-offers. Also, it only has time series and flex-offer data for the consumer's metering point and not (groups of) other metering points. For a node at a balance responsible party, on the other hand, it is necessary to represent both individual flex-offers from end-consumers and their aggregated flex-offers that are sent to the wholesale market. Further, it is necessary to represent information about the involved actors to know where the energy comes from, where it eventually gets consumed, who regulates the area, etc.

In a consumer node, some parts of the schema in Fig. 5 are thus not needed. For example, D_aggregationLevel is not needed and the attributes referencing it from D_timeSeries and F_flexOffer are also not needed. If they were present, they would always take the same values anyway and we thus say that they are *context-given*. Likewise representations of legal entities are context-given in a consumer node since the node only deals with a given consumer that belongs to a given market area etc. Further, a consumer node typically has limited computing resources and it can be beneficial to have a simpler database schema. At higher layers, there is typically much more computing power, but the data amounts may also be much bigger such that other schemas can be beneficial. To simplify the data management in a node, such as an end-consumer node, we employ actor-specific *specializations* of the schema. A specialized database schema S does not have to be able to represent all the data of the generic schema G, but may only be able to represent some of it and possibly in a modified form. S may have relations that are different from those in G and can, e.g., be a star schema. At an informal level, a specialized schema S can differ from the generic schema G in the following ways:

1. A new attribute a can be added to S if its values can be deterministically computed
   (a) from values of attributes in S or
   (b) from values of attributes in G *and* inverse functions that for each of these attributes in G can compute its value from the value of a are given.
2. An attribute from G can be left out from S if it in an instance of S always would take the same value if included (i.e., if it is context-given).
3. An attribute from G can be left out from S if we have a way of deterministically computing its value from the value of another attribute in S without knowing the state of G. In particular, a surrogate key is *not* enough to compute all other attributes of a relation.
4. An entire relation from G can be left out from S if all its attributes can be left out.
5. A relation in S can represent several relations from G that are equi-joined on foreign keys. A surrogate key used in a join may then be left out.

The data of an instance S must be obtainable from a number of queries on an instance of G such that the data for each relation in S is obtainable from one SPJ query. In particular, the queries may not use GROUP BY, HAVING, DISTINCT, UNION, INTERSECT, or EXCEPT from SQL. The queries can join relations on foreign keys, select an attribute once or leave it out if its values can be deduced from other included attributes or are context-given, and finally restrict the amount of tuples to those with certain values in certain attributes. The queries may not aggregate G data as this would prevent us from propagating modifications from the specialization instance back to the G instance.

6.1   Querying a Specialization

For a query q_S on a specialization S, it is possible to find a query q_G on the generic schema G that gives the same result: Since any relation in the specialization can be considered a view over one or several equi-joined relations in G, it is possible to find q_G from q_S by replacing each relation in q_S with its corresponding view definition over relations in G.

A query q_G on the generic schema G can under certain circumstances also be translated to a query q_S which gives the same result on a specialization S. Recall how a specialized schema can differ from the generic schema:

1. *A new attribute can be added if its value can be computed from values of other attributes.* Such attributes can be ignored since they obviously are not used by q_G.
2. *An attribute can be left out if it is context-given.* If q_G uses an attribute a that is context-given in S, q_S must use the appropriate constant instead of a.
3. *An attribute can be left out if it can be computed from another attribute in S.* If q_G uses an attribute b that is left out from S because it can be computed from another attribute c in S, q_S must do the necessary computation of b values by means of c, i.e., occurences of b should be replaced by $f(c)$ for a deterministic function f.
4. *A relation can be left out if it all its attributes can be left out.* If q_G uses such a relation r, all usages of attributes from r (which all necessarily are context-given or can be computed) can be replaced by appropriate constants.
5. *Several relations may have been equi-joined on foreign keys.* If a surrogate key has been left out from S, it cannot be used in queries on S. But since a surrogate key just is an integer with no special meaning, it would not make much sense to query for it anyway since it has already been used in a join to combine the right rows from two relations. We therefore assume that q_G does not query for a left-out surrogate key. Consider first the case where q_G equi-joins the relations $r_1, r_2, \ldots, r_n$ on foreign keys and S has the relation r' which holds the result of an equi-join of $r_1, r_2, \ldots, r_m$ $(m \le n)$ on foreign keys. In that case q_S can join r' and $r'_{m+1}, \ldots, r'_n$ where r'_i $(m + 1 \le i \le n)$ holds the corresponding data of r_i as found by applying these rules recursively. Now consider the case where S does not hold such an r', but instead holds a relation $\hat{r}$ with the result of an equi-join on foreign keys of $r_1, r_2, \ldots, r_N$ in G for an $N > n$. Then q_G in general cannot be transformed to a query on S that gives the same result since S might not represent all tuples from some r_j in G (in case no rows reference them) or represent some tuples too many times. As an example of the latter, if $n = 1$ in q_G and a relation in S holds the result of r_1 joined with r_2 (i.e., $N = 2$) on a foreign key from r_1 to r_2, tuples of r_2 might be represented several times in the resulting relation.

In other words, only item (5) can be a limitation. When one creates a specialization S, one should thus be aware that joining (i.e., denormalizing) too much makes some queries on the G schema impossible to translate to the S schema and get the same results. On the other hand this is not likely to be an issue in realistic settings. A designer would most likely not join G tables if, e.g., one of them holds rows that do not join with any rows from the other table(s) or if SELECTs from a particular table are an important query category.

A specialization can provide a simpler schema that fits the needs of a certain node and thus can be used instead of the general G schema. As discussed above, queries can always be translated from S to G and under certain circumstances from G to S. We, however, also wish to be able to do certain modifications on relations in S and be able to translate them to corresponding modifications on relations in G. We therefore now discuss which modification operations are allowed on data in a specialization instance.

6.2   Modifications

To maintain overall compatibility and consistency among nodes, it should be possible to propagate modifications made to S data back to G. Therefore, it is not all operations that are allowed in a specialization. Instead, any allowable operation on S that brings the database from a state s (obtained by applying the specialization's defining queries – or view definitions – on G in state g) to another state s' must be mappable to a number of operations on G that brings the database from g into a state g'. As discussed above, all relations in S can be seen as views on G. The state g' must then be such that if the view definitions, denoted V, are applied on a database with the schema G and the state g', the result is a database with the schema S and state s':

$$V(G, g') = (S, s')$$

For an attribute a added to S that also can be computed from other attributes X in S, we of course require that any modification to it is consistent. In other words, the value assigned to a should correspond to what can be computed from X. We now consider the possible modifications in turn and describe how they can be supported (if so) or why they cannot be supported.

Insertion is the most needed modification type for a specialization. A node with a specialization should be allowed to insert data about its own site, e.g., data about the energy consumption at the site. For a relation R_S in S which holds data from a single relation R_G in G, insertions can be supported in the following way. Some attributes of R_G may not be available in R_S, but they are then either context-given or computable from other attributes. Thus, for a row r_S inserted into R_S, we can find a corresponding row r_G to insert into R_G to achieve the state g'. This is similar to when views are updatable in SQL-92 [5] apart from that we do not get NULL values in left-out attributes but instead find proper values. Now consider a relation R_S which is the result of an equi-join of a sequence of relations $R_{G,1}, R_{G,2}, \ldots, R_{G,n}$ from G (possibly with some attributes left out if they can be computed or are context-given) where $R_{G,a}$ can have a foreign key referencing $R_{G,b}$ only if $a < b$. If a row r_S is inserted into R_S, we can for each $R_{G,i}$ (in the order $i = n, n-1, \ldots, 1$) find the corresponding part of the row r_S and add any computable or context-given attributes. We call this corresponding row part

$r_{G,i}$. If the state g of G is such that $r_{G,i}$ is not in $R_{G,i}$, it can be inserted (if the surrogate key is not present it should be added first and afterwards also added to all row parts for $R_{G,h}$ if $R_{G,h}$ references $R_{G,i}$). If it already is in $R_{G,i}$, nothing should be done. In SQL-92, insertions into "join views" (i.e., views with data from more than one relations) are not allowed. In SQL:1999 such insertions are sometimes allowed, but each of the view's columns should be uniquely traceable back to a single column in a single table [12]. We, on the other hand, allow a natural join where a column in R_S corresponds to two columns (namely, the primary key of $R_{G,b}$ and the foreign key of $R_{G,a}$) since we can consider one of the two columns as left-out due to computability. SQL does also allow the WITH CHECK OPTION to ensure that it is not possible to insert rows that would not appear in the view anyway (but it is not all RDBMSs that support it). This functionality is not available per se in a specialization, but would have to be emulated with CHECK constraints on the relations. The described method to support insertions does, however, not guarantee that a successful insertion into R_S can be mapped to successful insertions into the $R_{G,i}$ relations. For example, a primary key violation can occur when we try to insert into $R_{G,i}$ for some i. This would not be detected when inserting into R_S.

Deletion is a modification which rarely will be done in specializations. We anyway describe how it can be supported. For a row in R_S in S, we either have all corresponding G attributes directly available or can find them as argued above (possibly apart from surrogate keys). In the simple case where a row r_S is deleted from R_S which is not the result of a join, we can find the primary key value of the corresponding row r_G in R_G and use that to delete r_G. Consider for the more complex case again an R_S which is the result of an equi-join of a sequence of relations $R_{G,1}, R_{G,2}, \ldots, R_{G,n}$ from G (possibly with some attributes left out if they can be computed or are context-given) where $R_{G,a}$ can have a foreign key referencing $R_{G,b}$ only if $a < b$. We can either find the primary key value for each corresponding $r_{G,i}$ part or find values for all its other attributes and use them to identify the row to delete. Considering each of them in the order $i = 1, \ldots, n$, if the state g is such that $r_{G,i}$ is *not* referenced by any other row in any relation in G, it can be deleted. As with insertions, we are not guaranteed that a deletion in S results in one or more deletions in G. Another issue is whether we actually want a deletion in S to possibly result in deletions from more relations in G. Only the first deletion would delete detail data while the following ones would delete from the dimension hierarchy. It depends on the concrete case whether it makes most sense to delete from all corresponding G relations or only one, but the latter would often be what is wanted. It should thus be specified during the definition of a specialization how to handle deletions for relations that hold the results of equi-joined relations from G.

Update is simple to support in the case where R_S only has data from one relation, R_G, in G. In this case, the primary key value is again known and can be used to identify the corresponding row to update in R_G. For an R_S with data from several G relations, the situation is more complicated. With join views, a general problem is that some updates to the view cannot be mapped uniquely to a set of modifications to the base table [5]. In the case of specializations, we can, however, again benefit from knowing the primary key value for each corresponding row part and use this to do *upsertions*. Thus we can only allow updates to a relation holding the result of an equi-join of G relations if no surrogate keys have been left out. For an R_S which is the result of an equi-join of a sequence of relations $R_{G,1}, R_{G,2}, \ldots, R_{G,n}$ from G, we can for a row that is updated to r'_S consider the corresponding row parts in the order $n, \ldots, 1$. Assume that the primary key value for $R_{G,i}$ is p_i. The p_i values can be found for each corresponding row part of r'_S. If $R_{G,i}$ holds a row $r_{G,i}$ with the primary key value p_i, it should be updated to $r'_{G,i}$ (if $r_{G,i} \neq r'_{G,i}$). If $R_{G,i}$ does not hold such a row, $r'_{G,i}$ should be inserted. However, $r_{G,i}$ could be a corresponding row part of many rows in R_S, but if only one of these rows has been updated to go from the state s to the state s', we will not have that $V(G, g') = (S, s')$. To be able to map an update of R_S to G, we must therefore require that for any two rows x and y in R_S for which the primary key values of $x_{G,i}$ and $y_{G,i}$ are identical for some i, we also have that $x_{G,i} = y_{G,i}$ after the update. In other words, x and y should then have identical values for all attributes originating from R_G such that functional dependencies are maintained. This can be expensive to check and another and simpler solution is to only allow updates of the attributes in R_S originating from $R_{G,1}$.

Finally, we note that an attribute or relation that is left out from a specialization, obviously cannot be modified. If the context changes, e.g., if a consumer gets another energy supplier, a new S instance must be created for the new context.

In summary, insertions into a specialization are easy to support while deletions and updates are more complex. In particular, it is necessary to specify for a specialization how to handle deletions from a relation holding the result of a join of G relations. Also for updates, this should be specified. Updates do also require that no surrogate keys have been left out. In fact, only few modifications are expected to take place in a specialization. Typically, only insertions into one or few relations will be done; for example, meter readings from the node's location into F_timeSeriesInterval. A specialization definition can thus specify which relations may be modified and (for updates and deletions) how modifications to them should be mapped to modifications to G.

6.3 Examples

As a trivial example, it is possible to define a specialization of G with the same database schema. The mapping of modification operations is then simply the identity function. Another and more interesting specialization is for a prosumer C that has solar panels producing electricity and has agreed to buy all her remaining electricity from a given energy supplier. In this case, a specialization for a node at C's site does not need to represent the metering point (as it only has data for the single metering point) and the aggregation level (as no flex-offers are aggregated). Further, the energy type is always "solar energy" when produced by the prosumer and "undefined" when she buys electricity (the value of energyType can thus be computed from energyFlowDirection). Further, all flex-offers are offered by the consumer herself and only accepted by the single balance supplier and thus these values are context-given. Figure 7 shows S_C. Note how F_flexOffer and (several instances of) D_timeInterval have been joined leaving out the surrogate key timeIntervalId. New computed attributes with time stamps have, however, been added and they in turn allow the attributes of D_timeInterval to be left out. This is done to avoid the possibly expensive joins with D_timeInterval on the C node which is likely to have very limited hardware resources. D_timeSeries, D_typeEnergy, and D_type have also been joined as have F_timeSerieInterval and D_timeSeriesInterval. The attribute hour has also been added to the resulting relation. This attribute can be computed from time, but has been added to allow for efficient grouping when considering the hourly energy consumption. New time series intervals and flex-offers can be created at C's site and it must be possible to represent these in the generic schema as well. Thus insertions into D_timeSeries, F_timeSeriesInterval, and F_flexOffer are mapped to insertions into relations in G. Other modification operations are not allowed. If we wanted to also support updates of facts, we would have to include the left-out surrogate keys.

7 Queries

In this section, we give examples of interesting queries on data in MIRABEL DW. We first focus on queries on flex-offers and then on time series.

7.1 Queries on Flex-Offers

The first example, Q1, considers the flexibility in flex-offers, both with respect to time and amount of energy.

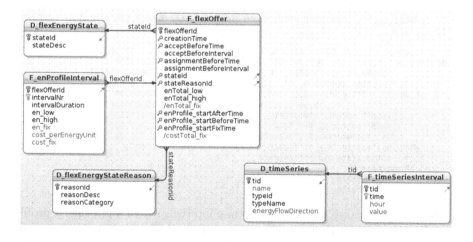

Fig. 7. The database schema S_C of a specialization

```
Q1:  SELECT AVG((enProfile_startBeforeTimeIntervalId -
                    enProfile_startAfterTimeIntervalId) *
              (SELECT SUM((en_high - en_low) * intervalDuration)
              FROM F_enProfileInterval i
              WHERE i.flexOfferId = f.flexOfferId)
              )
     FROM F_flexOffer f;
```

The query uses the flexibility with respect to time, i.e., the difference between when the flex-offer at the latest *has* to be executed and when it at the earliest *can* be scheduled. We assume that time interval IDs are assigned sequentially and thus use the difference between the IDs of the time intervals to find the flexibility. This flexibility is multiplied with the SUM of the energy flexibility in each profile interval. The energy flexibility in a profile interval is found as the length of the profile interval multiplied with the difference between the maximally required amount of energy and the minimally required amount of energy. Finally, the shown query considers the average of the combined flexibility for all flex-offers. The query is an example of a non-traditional kind of aggregation. If we consider a graph showing the relative start and end times for profile intervals on the X axis and the minimal and maximal energy amounts on the Y axis, the query Q1 finds the *area of energy flexibility* for all flex-offers and multiplies these with the length of their time flexibilities before the entire average is found. This number is primarily of interest *before* the scheduling gets done and a high number indicates much freedom in the scheduling while a low number shows that the considered flex-offers are not very flexible.

The next example, Q2, is of interest *after* the scheduling and gives the total amount of scheduled energy. This is a simple query which, however, must read data from many rows in a realistic setting (the DBMS we use does currently not support materialized views).

```
Q2: SELECT SUM(en_fix)
    FROM F_enProfileInterval;
```

Q3 is a more complex query to apply after scheduling has taken place. It builds a time series that, for each time interval ID, shows the amount of fixed energy.

```
Q3: SELECT timeIntervalId, SUM(en_fix_part)
    FROM (SELECT en_fix_part, ROW_NUMBER() OVER (PARTITION BY i.flexOfferId
                ORDER BY intervalNr) - 1 + f.enProfile_startFixTimeIntervalId
                AS timeIntervalId
         FROM (SELECT flexOfferId, intervalNr, en_fix / intervalDuration
                    AS en_fix_part, generate_series(1, intervalDuration)
               FROM F_enProfileInterval
               WHERE en_fix IS NOT NULL
              ) i, F_flexOffer f, D_flexEnergyState s
         WHERE i.flexOfferId = f.flexOfferId AND f.stateId = s.stateId
               AND s.stateDesc = 'Assigned'
         ) AS subquery
    GROUP BY timeIntervalId
    ORDER BY timeIntervalId;
```

The query computes the IDs of the time intervals where a flex-offer's profile intervals are executed. But a profile interval has a duration (in intervalDuration) which defines how many time intervals the profile interval spans. Therefore, it is necessary to (evenly) distribute the profile intervals' energy amounts over one or more time intervals. To do this, one "part" row is generated for each time interval a profile interval covers by means of generate_series. This happens in the innermost SELECT. The result of this is used by the second SELECT which also uses the SQL window function ROW_NUMBER to enumerate the rows in each partition where a partition consists of the part rows for a given flex offer and is ordered by the interval numbers. Thus, the resulting row number corresponds to the number of time intervals between the assigned start time for the entire flex offer and the part represented by the row (we subtract 1 since ROW_NUMBER counts from 1). When we add enProfile_startFixTimeInterval for the flex-offer, we get the ID of the absolute time interval when the part executes. Finally, the outermost SELECT aggregates the sums of fixed energy amounts over all parts belonging to a given time interval.

7.2 Queries on Time Series

Q4 is a query that finds the balance, i.e., the difference between produced and consumed energy, for a 24 hours period.

```
Q4: SELECT date, timeDesc,
          SUM(CASE energyFlowDirection WHEN 'Production' THEN value
                                       ELSE 0 END) AS production,
          SUM(CASE energyFlowDirection WHEN 'Consumption' THEN value
                                       ELSE 0 END) AS consumption
          SUM(CASE energyFlowDirection WHEN 'Production' THEN value
                                       WHEN 'Consumption' THEN -1 * value
                                       ELSE 0 END) AS balance
    FROM F_timeSeriesInterval f, D_timeSeries ts, D_type ty,
         D_typeEnergy te, D_timeInterval ti
    WHERE f.tid = ts.tid AND ts.typeId = ty.typeId AND te.energyTypeId =
          ty.typeId  AND ti.timeIntervalId = f.timeIntervalId AND
          te.energyOrigin = 'Metered' AND ti.date = '2011-06-01'
    GROUP BY ti.timeIntervalId
    ORDER BY ti.timeIntervalId;
```

The query Q4 is an example where we use the special attributes that only apply to some time series. In this example, we consider consumed and produced energy and we thus use energyFlowDirection and energyOrigin which only exist for energy time series. The query sums the production values, consumption values, and the difference between them for each time interval that belongs to a given date.

Our last example, Q5, is a query to find those time series where the average energy usage grouped on hours exceeds the average energy usage for the hour with 25 % or more at least 10 times.

```
Q5: WITH indavguse AS (
          SELECT tid, hour, COUNT(value) AS indcnt, AVG(value) AS indavg
          FROM F_timeSeriesInterval NATURAL JOIN D_timeInterval
          GROUP BY tid, hour
    ),
    totavguse AS (
          SELECT hour, SUM(indcnt * indavg) / SUM(indcnt) AS totavg
          FROM indavguse
          GROUP BY hour
```

```
),
overuse AS (
   SELECT tid, t.hour, indavg, totavg,
          COUNT(*) OVER (PARTITION BY tid) AS cnt
   FROM totavguse t, indavguse i
   WHERE t.hour = i.hour AND indavg >= 1.25 * totavg
)
SELECT tid, cnt, hour, indavg, totavg
FROM overuse
WHERE cnt > 10
ORDER BY tid, hour;
```

The query has Common Table Expressions (CTEs) in the WITH part. In the first CTE, `indavguse`, we compute a (temporary) table with the average hourly energy usage for each time series. The result is used again to compute the second CTE, `totavguse`, where we find the average energy use per hour among all time series (we could join F_timeSeriesInterval and D_timeInterval again, but it is faster to reuse the result of the previously computed CTE). In the third CTE, `overuse`, we join the the results of the two previous CTEs to find the IDs of time series and the hours from `indavguse` where the consumption is at least 25 % higher than the general hourly average consumption found in `totavguse`. Further, we use COUNT as a window function to count how many such hours we find for a given time series. Finally, we select the ID of the time series, the count of hours with an average energy usage at least 25 % higher than the average, and the consumption in the last SELECT clause.

8   Performance Study

In this section, we consider the queries from the previous section and use them to evaluate and compare the different MIRABEL DW schema alternatives presented in Sects. 2–6. This section is split into two parts. In the first part, we focus on the generalized variants of the MIRABEL DW schema – the *original* (unmodified, called "MDW"), *denormalized*, and *array-based* variants – and use them to compare the performance of the queries Q1–5. In the second part, we compare the original (MDW) generalized MIRABEL DW schema to a specialized variant (a specialization) by evaluating performance of Q4 in a resource-limited environment.

8.1   Performance of Q1–Q5 on the Generic Schemas

We now consider the queries Q1–Q5 on the described (original) schema of MIRABEL DW and its denormalized and array-based alternatives, denoted as "MDW", "denorm", and "array" respectively. In the denormalized variant, F_flexOffer and F_enProfileInterval are joined and so are F_timeSeriesInterval and D_timeSeries (however, with the name varchar attribute replaced by an integer to make it a typical fact table). In the array variant, the same tables are joined, but now grouped on all dimension references and with measures aggregated into arrays. For the tests, we use a real life data set with consumption data from 963 customers (the data originates from the MeRegio project [13]) and we synthetically generate flex-offers based on this data set. This gives rise to 963 (energy consumption) time series with 32.1 million time series values, and 3,1 million flex-offers. We test the performance on a Linux server with two Quad Core 1.86 GHz Intel Xeon CPUs, 16 GB RAM, 4 SATA 7200 RPM disks (with one dedicated to the DBMS). The DBMS is PostgreSQL 9.1 [15] and has the parameter shared_buffers set to 4 GB, temp_buffers to 128 MB, and work_mem to 96MB. All tables are "fully vacuumed" such that their disk representations only take up the needed space and do not occupy unused space. Further, the tables are "analyzed" such that their statistics are up-to-date. Each query is executed once in a warm-up round and then the queries are executed in a round-robin fashion such that each query gets executed five times. We report the average execution times. The results are shown in Fig. 8.

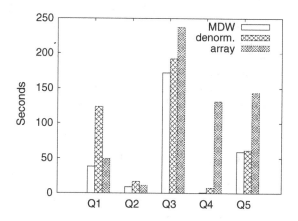

Fig. 8. Performance of Q1–Q5 on the generic schema

For **Q1**, it can be seen that the MDW variant is the fastest followed by the array variant (38.3 s and 49.1 s, respectively). These two query variants have similar plans, but with arrays there are fewer rows to process. On the other hand, these rows need to have their arrays "unnested" to produce as many values as there are rows to consider in the MDW variant. When the denormalized variant is considered, there are also many rows and these rows are wide. Further, the plan is not similar to the plans for the other variants as GROUP BY is necessary with this variant. This makes the denormalized variant the slowest (123.4 s).

For **Q2**, the MDW variant is again the fastest (8.9 s) to use. Again, the array variant is the second fastest (11.1 s). With this variant, the arrays must again be unnested to produce the values that are available in the rows in the MDW variant. The denormalized variant uses wider rows and is the slowest (16.8 s).

For **Q3**, the MDW variant remains the fastest (172.1 s) while the array variant now is the slowest (237.2 s) even though it avoids a join. On the other hand, the array variant requires a SELECT clause to unnest the array and an extra use of ROW_NUMBER to recreate the values from intervalNr which only are implicitly available from the array positions. The denormalized variant (192.2 s) is bit slower than the MDW variant even though it avoids a join.

For Q4, the MDW variant is significantly faster (0.8 s) than the others. The denormalized variant which avoids a join, uses an order of magnitude more time (7.7 s). The array variant is by far the slowest (131.9 s) as there is no index on timeIntervalId which is an array. Thus all rows must be processed and have their rows unnested to perform a join with D_timeInterval.

For Q5, the MDW and denormalized variants perform similarly (59.1 and 61.3 s, respectively). The queries involve the same number of rows and are identical apart from that the denormalized variant uses a wider table. For the array variant, the first CTE has to unnest two arrays and the query takes longer time (143.8 s).

To summarize, the MDW variant performs the best for all queries. Another interesting thing to consider, is the disk space usage. The tables F_flexOffer, F_enProfileInterval, F_timeSeriesInterval, and D_timeSeries take up 4.1 GB in the MDW variant (not counting indexes). Their alternative representations take up 7.0 GB in the denormalized variant and 1.9 GB in the array variant, respectively. It notable how little space the array variant uses compared to the other variants due to its fewer number of rows (and thus fewer space-consuming row headers). Overall, the MDW variant is a good choice considering both its performance and space requirements.

8.2   Performance of Q4 on the Specialized and Generic Schemas

We now consider MIRABEL DW at a prosumer node (e.g., a smart-meter), which uses MIRABEL DW for storing, among other entities, electricity consumption and production measurements. As this node is expected to have limited computing and storage capabilities, we consider a MIRABEL DW specialization as opposed to the full MIRABEL DW schema for the storage of measurements. To simulate a resource-limited environment, we use three instances of the Oracle VirtualBox virtual machine (VM), each of which runs the lightweight Linux DSL 4.2.5 OS and the SQLite 3.3.10 DBMS. We deploy these instances on the machine from Sect. 8.1. The configurations of these VM instances are as follows:

VM(100,1024). The CPU clock speed/frequency is 100 % of the host machine, but the memory (RAM) is limited to 1024 MB.
VM(100,12). The CPU clock speed/frequency is 100 % of the host machine, but the memory (RAM) is limited to 12 MB.
VM(10,12). The CPU clock speed/frequency is limited to 10 % of the host machine, and the memory (RAM) is limited to 12 MB.

For the experiment, we use a dataset with consumption and production measurements collected every 15 min within an eight year time interval. These are stored as two separate time series in two databases – the first database in the generic MIRABEL DW schema G (MDW) and the second database in the specialization schema S_C from Sect. 6.3. By varying the total amount of

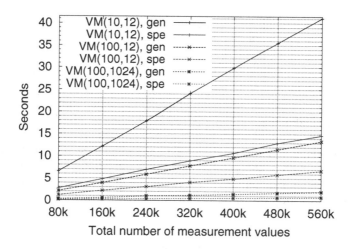

Fig. 9. Performance of Q4 on the specialized and generic schemas

measurements stored in MIRABEL DW, we submit (variants of the) query Q4 for computing the difference between production and consumption (balance) for a 24 h period for a selected day, and measure the total time required to evaluate Q4 on each of these three VM instances. The results of this experiment are shown in Fig. 9.

As seen in the figure, the query Q4 takes up to 2.5 times more time to evaluate for the generic schema in comparison with the specialized schema. The fewer resources the node has, the more it pays of to use the specialization. In summary, we can see that the use of specialized schemas has a big potential for resource-constrained devices such a smart meters.

9 Related Work

In the energy sector, there is a number of standardized data models used to represent the major objects in an electric utility enterprise [8] as well as to define administrative data internally interchanged between European electricity markets [6,7]. These models focus on various aspects of energy trading and physical electricity delivery, and specify 1) components of a power system at the electrical level, 2) actors and roles involved in the energy trading, 3) relationships and data exchange between those entities. These models are used as a basis for the MIRACLE data model [10], which further enriches them with the concept of shiftable consumption and production. All these models, however, focus on a semantic rather than the storage or the management of energy-related entities. By focusing on two most important entities in MIRABEL, i.e., time series and flex-offers, this paper, on the other hand, presents data representation models for these two types of entities offering a convenient storage and a good performance of analytical queries.

This paper is a significant extension of a previous conference paper [17], and the papers are the first to deal with the storage of flex-offers. There are previous works which focus on time series and warehousing, e.g. UML-based modeling of time-series in DWs [20], and temporal aggregation of multidimensional data [3], and temporal DWs exploiting research results from the field of temporal databases [11]. Our modeling of different time-series types has similarities with Bauer et al.'s work [1]. They discuss "locally valid dimensional attributes" whose existence depends on values of dimensional elements. This is the case, e.g., for our attribute energyType which only exists if the D_type value represents an energy time-series. The problem of representing all these attributes in a single dimension table (as in a typical star schema) is that there will be many NULLs in the held data. Bauer et al. propose to have separate tables with the specific attributes and then create views "on top" of these with common attributes as well as textual values showing the name of the relation the data comes from which can be used for hierarchical classification. In contrast, we use tables (and not views) for the common attributes of a dimension and then represent special attributes that only exist for some dimensional values in other tables that reference the table with the common attributes. This makes it possible to declare foreign keys to the dimension table with the common attributes and also declare indexes and constraints on these tables. Bauer et al. also propose to use table inheritance to represent such cases. This would also be possible in our DBMS [15], but constraints cannot be enforced on child tables then. Yu et al. [19] present an approach for storing very big time series from sensor networks using Cloud technologies such as Bigtable [4]. In contrast, we use relational storage technology and further our individual DWs cannot be merged into a single, shared DW due to privacy concerns, as the data comes from many competing companies.

In the current paper, we consider different representations of profile intervals and time series intervals which can be considered as facts with multi-valued measures. The latter case also has a many-many relationship between the time series facts and the time interval dimension. Previous work [18] has considered many-many relationships between fact tables and dimension tables. Our denormalized representation is similar to one of the methods of [18] whereas our other approaches with fact tables referencing other fact tables and measure values in arrays, respectively, are different.

Compared to the conference version [17], the current paper also provides descriptions of how to allow specializations of the generic schema. This includes how to translate queries between them and how to support modifications to the data in a specialization such that the generic schema can be modified correspondingly. The latter is related to updatable views as discussed, e.g., in [5,12], and includes a range of new experiments that compare standard and specialized schemas.

10   Conclusion

In this paper, we have presented a unified, generic DW schema for managing the complex energy data in a smart grid, including actors playing roles, flex-offers, and different types of time series. The schema has a number of interesting complexities such as facts about facts and composed non-atomic facts. The different nodes will hold different parts of the data accordingly to the roles of the node owners and the data will be at different aggregation levels at different nodes. The same schema can, however, be used for all kinds of nodes. We have considered different alternatives for the schema modeling using denormalization and arrays, respectively, but based on the performance and space usage, the chosen design is favourable. Further, we have described how to allow specialized versions of the schema for different types of nodes, but such that queries can be formulated on the generic schema and automatically be translated to the specialized schemas (and vice versa) to make the results combinable. We also described how to support modifications on specializations.

In the near future, we are going to extend the DW schema to be able to handle other energy-specific entities such as operating schedules, parameters, and power network constraints, statuses, loads, and spatial models. Furthermore, we will perform large-scale simulations with realistic data amounts from different types of nodes. We will also perform large-scale simulations using nodes that use role-specific specializations of the general DW schema. As part of it, we will simulate the update propagation between different specializations.

References

1. Bauer, A., Hümmer, W., Lehner, W.: An alternative relational OLAP modeling approach. In: Kambayashi, Y., Mohania, M., Tjoa, A.M. (eds.) DaWaK 2000. LNCS, vol. 1874, p. 189. Springer, Heidelberg (2000)
2. Boehm, M., et al.: Data management in the MIRABEL smart grid system. In: Proceedings of EDBT/ICDT Workshops (2012)
3. Böhlen, M.H., Gamper, J., Jensen, C.S.: Multi-dimensional aggregation for temporal data. In: Ioannidis, Y., Scholl, M.H., Schmidt, J.W., Matthes, F., Hatzopoulos, M., Böhm, K., Kemper, A., Grust, T., Böhm, C. (eds.) EDBT 2006. LNCS, vol. 3896, pp. 257–275. Springer, Heidelberg (2006)
4. Chang, F., et al.: Bigtable: a distributed storage system for structured data. TOCS **26**(2), 205–218 (2008)
5. Elmasri, R., Navathe, S.B.: Fundamental of Database Systems, 4th edn. Addison Wesley, Boston (2004)
6. European Network of Transmission System Operators for Electricity. The Harmonised Electricity Market Role Model, version 2011–01. www.ebix.org/Documents/role_model_v2011_01.pdf as of 03 March 2014
7. Introduction to Business Requirements and Information Models. www.ebix.org/documents/Introduction%20to%20ebIX%20Models%200.0.D.pdf as of 03 March 2014
8. IEC61970-301 Ed. 2, Energy management system application program interface (EMS-API) - Part 301: Common information model (CIM) base, International Electrotechnical Commission (2009)
9. Jensen, C.S., Pedersen, T.B., Thomsen, C.: Multidimensional Databases and Data Warehousing. Morgan & Claypool, San Rafael (2010)

10. Konsman, M.J., Rumph, F.J.: MIRABEL Deliverable 2.3: Final data model, specification of request and negotiation messages and contracts. https://www.db.inf. tu-dresden.de/miracle/publications/D2.3.pdf as of 03 March 2014
11. Malinowski, E., Zimányi, E.: Advanced Data Warehouse Design From Conventional to Spatial and Temporal Applications. Springer, Heidelberg (2009)
12. Melton, J., Simon, A.R.: SQL:1999 Understanding Relational Language Components. Morgan Kaufmann, Boston (2001)
13. www.meregio.de/en/ as of 03 March 2014
14. As of 03 March 2014. www.mirabel-project.eu/
15. www.postgresql.org as of 03 March 2014
16. Šikšnys, L., Khalefa, M.E., Pedersen, T.B.: Aggregating and disaggregating flexibility objects. In: Ailamaki, A., Bowers, S. (eds.) SSDBM 2012. LNCS, vol. 7338, pp. 379–396. Springer, Heidelberg (2012)
17. Siksnys, L., Thomsen, C., Pedersen, T.B.: MIRABEL DW: managing complex energy data in a smart grid. In: Cuzzocrea, A., Dayal, U. (eds.) DaWaK 2012. LNCS, vol. 7448, pp. 443–457. Springer, Heidelberg (2012)
18. Song, I-Y., et al.: An analysis of many-to-many relationships between fact and dimension tables in dimensional modeling. In: Proceedings of DMDW (2001)
19. Yu, B., et al.: On managing very large senor-network data using bigtable. In: Proceedings of CCGrid (2012)
20. Zubcoff, J., Pardillo, J., Trujillo, J.: A UML profile for the conceptual modelling of data-mining with time-series in data warehouses. Inf. Softw. Technol. 51(6), 977–992 (2008)

Modular Neural Networks for Extending OLAP to Prediction

Wiem Abdelbaki[1,2]([⊠]), Sadok Ben Yahia[1], and Riadh Ben Messaoud[3]

[1] Faculty of Sciences of Tunis, University of Tunis El-Manar, LIPAH-LR 11ES14,
2092 Tunis, Tunisia
sadok.benyahia@fst.rnu.tn
[2] Department of Information Systems,
College of Economics Management and Information Systems, University of Nizwa,
616 Nizwa, Nizwa, Sultanate of Oman
wiem.abdelbaki@gmail.com
[3] Faculty of Economics and Management of Nabeul, University of Carthage,
8000 Nabeul, Tunisia
riadh.benmessaoud@fsegn.rnu.tn

Abstract. On-line Analytical Processing (OLAP) represents a good applications package to explore and navigate into data cubes. Though, it is limited to exploratory tasks. It does not assist the decision maker in performing information investigation. Thus, various studies have been trying to extend OLAP to new capabilities by coupling it with data mining algorithms.

Our current proposal stands within this trend. It has two major contributions. First, a Multi-perspectives Cube Exploration Framework (MCEF) is introduced. It is a generalized framework designed to assist the application of classical data mining algorithm on OLAP cubes. Second, a Neural Approach for Prediction over High-dimensional Cubes (NAP-HC) is also introduced, which extends Modular Neural Networks (MNN)s architecture to multidimensional context of OLAP cubes, to predict non-existent measures. A preprocessing stage is embedded in NAP-HC to assist it in facing up the challenges arising from the particularity of OLAP cubes. It consists of an OLAP oriented cube exploration strategy coupled with a dimensions reduction step that reposes on the Principal Component Analysis (PCA). Carried out experiments highlight the efficiency of MCEF in assisting the application of MNNs on OLAP cubes and the high predictive capabilities of NAP-HC.

Keywords: Data warehouse · OLAP · Data mining · Principal Component Analysis · Multilayer Perceptrons · Modular Neural Networks

1 Introduction

Data warehouses are the corner stone in the Business Intelligence (BI) roadmap. They are used to store analysis contexts within multidimensional data structures referred to as *Data Cubes* [1]. They are usually manipulated through On-line Analytical Processing (OLAP) applications to enable senior managers exploring information and getting BI reportings through interactive dashboards.

© Springer-Verlag Berlin Heidelberg 2015
A. Hameurlain et al. (Eds.): TLDKS XXI, LNCS 9260, pp. 73–93, 2015.
DOI: 10.1007/978-3-662-47804-2_4

Needless to mention that OLAP tools provide efficient solutions to navigate through data cubes. However, it is restricted to exploration tasks. Goil and Choudhary argue that coupling OLAP with data mining techniques increases its efficiency [2], and enables it to assist decision makers in performing advanced knowledge discovery tasks. Since then, several studies put the focus on enhancing OLAP by coupling it with data mining techniques to respond to various analysis purposes, e.g. cube exploration [3] and association rule mining [4].

Nevertheless, despite the fact that, data warehouses should fundamentally contain integrated data [1], generally, data cubes exploration discloses sparse structures within several empty measures. In this respect, empty measures correspond to non-existent facts, reflecting either out-of-date events that did not happen, or future events that have not yet occurred and may happen in the future. Empty measures represent a source of frustration for the enterprise management, especially when strategic decisions need to be taken.

Predicting non-existent measures would consolidate BI reporting. It would even provide new opportunities to BI analysts by enlarging their dashboard picture and empowering them with knowledge on what may occur if non-existent facts had already happened. For instance, it will be very useful to a car Sale Company to predict the potential turnover that a new agency could produce in a new city by the end of next year. This indicator will definitely help the company's management to assess the potential investment.

Despite the fundamental Cood's statement of goal seeking analysis models (such as "What if" analysis) required in OLAP applications since the early 90's [5], most of the recent OLAP products still lack an effective implementation of this feature. Recently, new approaches have been attempting to extend OLAP to prediction capabilities [6,7]. However, to the best of our knowledge; none of them provides BI analysts with explicit values of non-existent measures.

The current work fits within the approaches trying to extend OLAP to advanced abilities by coupling it with data mining techniques. It introduces two main contributions. The first one consists of a novel generalized framework, called *Multi-perspectives Cube Exploration Framework* (MCEF). It is designed to enable the application of classical data mining techniques on OLAP cubes. As for the second contribution, it consists of a measure prediction technique, called *Neural Approach for Prediction over High-dimensional Cubes* (NAP-HC). It is based on Modular Neural Networks (MNN)s and designed under the MCEF formalism.

This paper is organized as follows. In Sect. 2, we expose a state of the art of works related to predictions in data cubes. We introduce and formalize the MCEF in Sect. 3. Section 4 details the formalization of NAP-HC. In Sect. 5, we carry out experiments investigating the effectiveness and the efficiency of our proposals. Finally, Sect. 6 summarizes our contributions and addresses future research directions.

2   Related Work

In recent years, several studies have been addressing the issue of extending OLAP to advanced analysis capacities. They were driven under different motivations

Table 1. Proposals addressing prediction in data cubes

Proposal	Goal	Optimization	Reduction	Measures	Values
Sarawagi et al. [3]	Exploration	+	−	−	−
Palpanas et al. [8]	Compression	−	+	−	−
Chen et al. [9]	Prediction	+	−	+	−
Cuzzocrea [10]	Query approximation	+	+	+	+
Chen et al. [11]	Compression	−	+	+	−
Bodin-Niemczuk et al. [6]	Prediction	+	−	−	−
Cuzzocrea and Saccà [12]	Privacy preserving	+	−	−	+
Agarwal and Chen [7]	Prediction	+	−	+	−
Our approach	Prediction	−	+	+	+

e.g. discovery-driven cube exploration [3], association rules mining [13], cube compression [11]. Thus, they are based on various concepts and methodologies. In this section, we focus on those having a close linkage with prediction in data warehouses.

Table 1 summarizes the proposals attempting to extend OLAP to prediction. These proposals are detailed according to five main criteria: (1) What is the overall goal of the proposal? (2) Does the proposal include an algorithmic optimization? (3) Does it use a reduction technique? (4) Does it introduce new classes of measures? And (5) Does it provide explicit predicted values of empty measures? We note (+) if the proposal fulfils the criteria, and (−) if in the opposite situations.

Sarawagi et al. proposed to assist data warehouse users when exploring data by detecting exceptions [3]. Their approach is based on a log-linear model. Palpanas et al. used the principle of information entropy to build a probabilistic model capable of detecting measure deviations [8]. To compress data cubes, Chen et al. introduced the concept of *Prediction cubes*, where the score or the probabilities of measures are fetched beside their original values [9]. Prediction Cubes are exploited to build prediction models, which predict low-level measures from high-level pre-calculated aggregates. In [10], Cuzzocrea propose a statistical framework that provides probabilistic bounds on approximate answers. This framework's main goal consists at supporting OLAP applications in overcoming queries' answering, which are considered among the main bottleneck for of OLAP applications. More specifically, it aims at enhancing the accuracy of the approximate answers. To do so, the framework reposes on a sampling technique, which ensures the quality of the approximate answers and generates the probabilistic guarantees on their approximation's degree. On the other hand, to ensure the scalability of the proposal, the author extends it with a previously proposed data cube dimensions reduction technique [14], based on the Karhunen-Loeve transform [15]. In [11], Chen *et al.* proposed a new type of multidimensional structures called *Regression Cubes*, which contain compressible measures. Regression cube cells indicate measure variations and tendency. Cuzzocrea and Saccà address the computation of privacy preserving OLAP aggregations [12]. Their framework reposes on sampling-based data cube compression. Its strength consists in the fact that the generated privacy preserving aggregates stills allow the evaluation of approximate answers. Agarwal and Chen introduced a new data cube class called *Latent-Variable Cube*, built over a statistical model [7]. It enables the computation of aggregate functions, such as mean and variance over latent variables. Bodin-Niemczuk *et al.* propose to equip OLAP with a regression tree to predict measures of forthcoming facts [6].

Most of the cited proposals recognize that the combination of the important dimensionality and huge volumes of data cubes represent a serious challenge for most of the approaches trying to apply a data mining technique on OLAP cubes. To face up this challenge, some proposals consider a preprocessing stage to reduce the dimensionality effect on algorithm's performance [8,10,11], while some others rather rely on heuristics to optimize implemented algorithms [3,7,9,12]. In our case, we include a PCA-based preprocessing stage in our prediction proposal, which reduces the data cubes dimensionality and generates concentrated, information preserving training sets for the prediction stage.

We notice that all the approaches having different goals than measure's prediction do not provide explicit values for measures. While [3,6,8,10,12] provide approximations of non-existent measures, [7,9,11] introduce new classes of data cubes within new measures generated over the existing ones. Nevertheless, this is totally justified since most of the cited proposals largely meet their main objectives. Among them all, only Bodin-Niemczuk *et al.*'s proposal shares the particular goal of non-existent measure prediction with us [6]. However, the output of this approach is a set of discretized values of the targeted measures. Subsequently, the relevance of the results is strongly depending of the nature and the range of the produced intervals. This issue may not satisfy the analyst who aims an explicit precise decision. The predictive model that we introduce in this paper provides the decision maker with explicit predicted values of non-existent measures, which do not require any further processing.

3   Multi-Perspectives Cube Exploration Framework

3.1   Motivations

Following the success of data warehouse technology, OLAP tools, which are limited to exploratory tasks, are no longer sufficient to meet the increasing needs of OLAP users. Thus, several approaches have been trying to extend OLAP to new abilities by coupling it with data mining techniques to deal with different issues, e.g. cube exploration [3], association rules mining [13], non-existent measures prediction [16,17]. However, most of these proposals consist at specialized solutions, which are tightly related to their particular goals. Thus, even if most

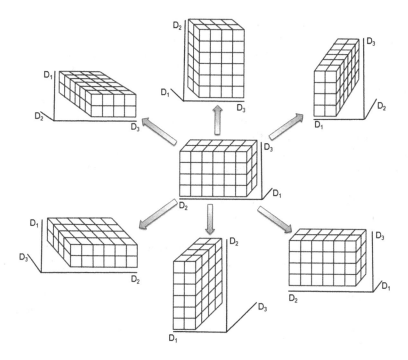

Fig. 1. Possible views of a 3 dimensions cube

of them share the same general motivation of applying data mining algorithms on OLAP cubes, they cannot exploit the already designed formalism of each other's.

We believe that a uniform standardized framework that assists the extension classical data mining algorithms to OLAP cubes' context could turn out to be very useful. Firstly, it offers a uniform ready-to-deploy formalism for the forthcoming proposals aiming to extend OLAP cubes context with mining algorithms. Secondly, and most importantly, it opens the doors for interoperability between the different proposals, since they will be based on the same formalism and handling the same components type. Doing do, the outputs of one proposal could be exploited as the inputs of another. For example, an analysis could start with cube exploration [3], passes through association rules extraction [13] and ends by non-existent measures prediction [16]. Doing so, the analysis would end-up with more efficient reportings. Furthermore, implementing this platform opens the doors for producing software packages that include multiple cube mining algorithms, similarly to *Weka* and *Tanagra* packages [18], which are dedicated to the bi-dimensional context.

On the other hand, the application of cube mining techniques is generally preceded by an in-depth analysis step, which consists of a vertical cube exploration that ends by selecting the most suitable hierarchical levels for the analysis. However, even if it is commonly ignored, horizontal cube exploration, which consists at the selection of the most convenient dimensions' distribution over cube axes for the analysis, have to be considered.

Actually, each dimensions' distribution across the cube axes generates a different cube view, i.e. data presentation. As illustrated in Fig. 1, multiple data presentations could be obtained from a single three-dimensional data cube, following the dimensions' distribution across its axes. Cuzzocrea and Mansmann state that the efficiency of data representation has an important impact on the data exploration and visualization [19].

Actually, the dimensions' distribution defines the way data is delivered to the data mining algorithm. Therefore, it has a great impact on multiple data mining techniques, especially, the ones that are sensitive to the way data is delivered to them. For these techniques, considering a single dimensions' distribution may promote some dimensions at the expense of others, which causes the loss of the relevant patterns that could be generated over the unexplored views. Nevertheless, most of the researches ignore horizontal cube exploration and limit their analysis to a single dimensions' distribution, usually, implicitly, selected following the user's preferences.

In [20], Ramakrishnan and Chen highlight that mining large datasets requires a principled way to explore the large space of possibilities and alternatives. We further claim that; in order to obtain representative results from rich versatile structures such as data cubes, horizontal cube exploration should be addressed and reinforced.

To concretize these considerations, we design a Multi-perspectives Cube Exploration Framework (MCEF). It is a generalized framework that assists the application of classical data mining algorithm on OLAP cubes, while supporting both horizontal and vertical cube explorations, designed to meet the following goals:

1. Supporting the application of classical data mining algorithms on OLAP cubes;
2. Considering both horizontal and vertical data cubes exploration approaches;
3. Ensuring equitable contributions of the dimensions to the analysis;
4. Preserving the semantics linking members to their respective dimensions and to other dimensions' members;
5. Covering all the possible measures' variations in terms of dimensions' distributions;
6. Enabling BI analyst to define customized analysis contexts.

3.2   Multi-Perspectives Cube Exploration Framework

In this subsection, we thoroughly describe and elaborate the MCEF formalism.

We start by recalling Ben Messaoud et al. data cube definitions, which we intend to reuse [13]. Afterwards, we introduce the new definitions required to develop the MCEF formalism.

We start by recalling [13] data cube definitions, which we reuse in MCEF formalization. Let $\mathcal{C}$ be a data cube having the following properties:

- $\mathcal{C}$ has a nonempty set of d dimensions $\mathcal{D} = \{D_i\}_{(1 \leq i \leq d)}$;
- $\mathcal{C}$ contains a nonempty set of m measures $\mathcal{M} = \{M_q\}_{(1 \leq q \leq m)}$;
- $\mathcal{H}_i$ is the set of hierarchical levels of the dimension D_i. $H_j^i \in \mathcal{H}_i$ is the jth hierarchical level of D_i. In Fig. 2, H_1^2 of D_2 is $Product\_name$.
- $\mathcal{A}_{ij}$ is the set of members of the hierarchical level H_j^i; $\theta_t^{ij} \in \mathcal{A}_{ij}$ is the tth member of the jth hierarchical level of the dimension D_i. In Fig. 2, θ_5^{22} is iPod.

Definition 1 Inter-dimensional predicate. *Let $\mathcal{D}_a \in \mathcal{D}$ be a nonempty set of p dimensions $\{D_1, ..., D_p\}_{(1 \leq p \leq d)}$ from the data cube $\mathcal{C}$. An inter-dimensional-predicate defines a conjunction of non-repetitive members, i.e., each dimension has a distinct member in the expression. $\Theta^a = (\theta_t^{mi} \wedge ... \wedge \theta_s^{nj})$ is called an inter-dimensional predicate in $\mathcal{D}_a$ if θ_t^{mi} is the tth member of the ith hierarchical level of the dimension D_m and θ_s^{nj} is the sth member of the jth hierarchical level of the dimension D_n, and $\{D_m, D_n\} \in \mathcal{D}_a$.*

In Fig. 2, let $\mathcal{D}_a = \{D_1, D_2\}$ be a set of dimensions of $\mathcal{C}$, a random inter-dimensional predicate Θ^a can of be of the form: $(\langle \theta_1^{11} \in \mathcal{A}_{11} \rangle \wedge \langle \theta_5^{22} \in \mathcal{A}_{22} \rangle)$, e.g. $(\langle quarter1 \rangle \wedge \langle iPod \rangle)$.

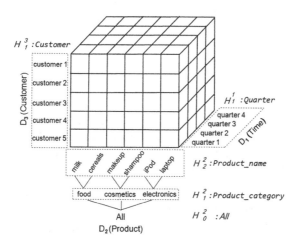

Fig. 2. Example of a data cube

Hereafter, we define the new concepts required to formalize our proposal.

Definition 2 Inter-dimensional hierarchical predicate. *Let $\mathcal{D}_a \in \mathcal{D}$ be a nonempty set of p dimensions $\{D_1, ..., D_p\}_{(1 \leq p \leq d)}$ from the data cube $\mathcal{C}$. An inter-dimensional hierarchical predicate defines a conjunction of distinct hierarchical levels of non-repetitive dimensions. $\Omega^a = (H_m^s \wedge ... \wedge H_n^t)$ is called an inter-dimensional hierarchical predicate of $\mathcal{D}_a$ if H_m^s is the mth hierarchical level in D_s, H_n^t is the nth hierarchical level in D_t and $D_s \neq D_t$.*

In Fig. 2, let $\mathcal{D}_a = \{D_1, D_2\}$ be a set of dimensions of the data cube $\mathcal{C}$. $\Omega_i^a = (\langle H_1^1 \in \mathcal{H}_1 \rangle \wedge \langle H_2^2 \in \mathcal{H}_2 \rangle)$, which is, $(\langle \texttt{Quarter} \rangle \wedge \langle \texttt{Product\_name} \rangle)$ is a random inter-dimensional hierarchical predicate of $\mathcal{D}_a$.

In the sequel of this paper, let $\mathcal{D}_c = \{D_1, \ldots, D_c\}_{(0 \leq c \leq d-2)}$,

$\mathcal{D}_v = \{D_1, \ldots, D_v\}_{(0 \leq v \leq d-2)}$ and $\mathcal{D}_r = \{D_1, \ldots, D_r\}_{(0 \leq r \leq d-2)}$ be three non-empty sets of c, v and r distinct dimensions, respectively; with $c + v + r \leq d$ and Ω^c, Ω^v, Ω^r be three inter-dimensional hierarchical predicates of $\mathcal{D}_c$, $\mathcal{D}_v$ and $\mathcal{D}_r$, respectively.

Let $\Theta^c, \Theta^v, \Theta^r$ be three inter-dimensional predicates in $\mathcal{D}_c, \mathcal{D}_v, \mathcal{D}_r$, respectively, and let $\Omega^c, \Omega^v, \Omega^r$ be three inter-dimensional hierarchical predicates of $\mathcal{D}_c, \mathcal{D}_v, \mathcal{D}_r$, respectively.

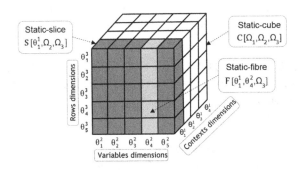

Fig. 3. Static-cube, static-slice and static-fibre

Definition 3 Static-cube. *We denote by $C[\Omega^c, \Omega^v, \Omega^r]$ a static-cube of a data cube $\mathcal{C}$. It is the fixed distribution of the cells obtained by the application of the OLAP Dice operator on $\mathcal{C}$, following, Ω^c, Ω^v and Ω^r . $\mathcal{C}$ is identifiable by the distribution of $\Omega^c, \Omega^v, \Omega^r$, across $\mathcal{C}$ axes.*

The dimensions of C are distributed over three classes; *Contexts* dimensions $\mathcal{D}_c$, *Variables* dimensions $\mathcal{D}_v$ and *Rows* dimensions $\mathcal{D}_r$, which we refer to as MCEF dimensions classes. Each of these classes is designed to ensure a particular role:

- **Contexts dimensions :** The set of attributes generated over these dimensions combination serves in identifying the different analysis subcontexts
- **Variables dimensions :** The set of attributes generated over these dimensions combination is considered as a set of variables.
- **Rows dimensions :** The set of attributes generated over these dimensions combination is considered as a set of observations.

The main goal of the static-cube concept is to depict all the possible dimensions' distributions across the cube axes. An illustrative example is shown in Fig. 3. The latter represents the static-cube $C[\Omega^1, \Omega^2, \Omega^3]$, with $\mathcal{D}_c = \{D_1\}$, $\mathcal{D}_v = \{D_2\}$, $\mathcal{D}_r = \{D_3\}$ as the sets of Contexts, Variables and Rows dimensions, respectively.

Following the BI analyst preference, an analysis could either consider the most relevant static-cube to the analysis or involve the entire set of static-cubes.

On the other hand, a single three-dimensional OLAP cube's slice can generate two distinct bi-dimensional tables, with one representing the transpose of the other. This might cause the confusion of the data mining technique and lead to inconsistent results. To solve this issue, in what follows, we introduce the concept of static-slice, which enables the distinction between the different bi-dimensional tables that can be generated over a single OLAP slice.

Definition 4 *Static-slice. We denote by $S[\Theta^c, \Omega^v, \Omega^r]$ a static-slice of a static-cube C. It is the fixed distribution of the cells obtained by the application of the OLAP Slice operator on C, following Θ^c, Ω^v and Ω^r. S have the same MCEF dimensions distribution of C.*

The concept of *static-slice* is designed to enable browsing static-cubes in a principled way, following the different MCEF classes. For instance, the dark grey coloured cells in Fig. 3 represent the static-slice $S[\Theta_1^1, \Omega^2, \Omega^3]$.

Definition 5 *Static-fibre. We denote by $F[\Theta^c, \Theta^v, \Omega^r]$ a static-fibre of a static-slice S. It is the fixed distribution of the cells obtained by the application of the OLAP Dice operator on C, following Θ^c, Θ^v and Ω^r. F have the same MCEF dimensions distribution of S.*

The concept of *static-fibre* is designed to enable browsing static-slices in a principled way, following the different MCEF classes. As instance, the light grey coloured cells in Fig. 3 represent the static-fibre $F[\Theta_1^1, \Theta_4^2, \Omega^3]$.

Data mining could be classified into two distinct categories. The first one concerns the data mining techniques that are not sensitive to the way data is provided to them. Therefore, they would generate the same mining outcome with the different static-cube. As for the second category, it concerns the data mining that are sensitive to the way data is provided to them, which makes each static-cube a unique dataset. For this type of category, the most optimal scenario that ensures equitability between dimensions consists in involving all the potential static-cubes in the analysis. Then, following the analysis aims, the BI analyst can either combine the obtained results or consider them separately. Still, this solution is very expensive and may turn to be non-effective, especially if in the case of online deployment. The other alternative consists in limiting the analysis to a the most relevant the static-cubes.

4   Neural Approach for Prediction over High-Dimensional Cubes

4.1   Overview

As far as we know, despite their proven performances, Neural Networks (NN)s are not yet exploited in OLAP cubes' context. This is due to multiple factors. First,

NNs generalization capabilities become limited when handling high-dimensional datasets. Second, the computational requirements of NNs increase drastically with the increase of inputs' number, which slows down the learning rates [21]. Third, highly correlated data may corrupt the training phase of NNs and degrade their generalization capability [22].

On the other hand, Modular Neural Networks (MNN)s represent a well-established technique in the field of machine learning. They are generally composed by set of (NN)s called *modules* and a *combiner* system [23]. They are based on the "divide and conquer" principle. They undertake a complex problem, divide it into smaller tasks and distribute them over the modules. Modules can be trained independently or sequentially targeting the same task. While, the combiner system processes their outputs to generate a conclusive analysis result of the entire system.

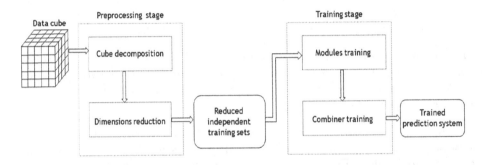

Fig. 4. Overview of NAP-HC architecture

Multiple researches claim that MNNs overcome multiple limitations of single NNs [23–25]. Melin and Castillo state that MNNs are very effective to overcome the problems defined over high-dimensional space and having high complexity [25]. Happel and Murre [26] state that MNNs enable the application of NNs on large-scale data [26]. Gallinari claim that MNNs reduce the model complexity, provide robustness and enable data sources fusion [24]. Sharkey sheds the light on the fact that decomposing a large complex task into modular components makes the system easier to understand and to modify [23]. These factors, make MNNs highly promising candidates to overcome the limitations of NNs with multidimensional large structures, such as data cubes.

Despite the fact that MNNs might resolve multiple problems related to the application on NNs on OLAP cubes, the high dimensionality and the correlated measures still represent thriving challenges that could deteriorate the training process quality. Recently, some studies have been interested in Principal Component Analysis (PCA) to reduce the dimensionality of prediction models inputs [27,28]. The PCA is an exploratory statistical procedure, which aims at transforming the original correlated variables into a smaller set of uncorrelated ones, called *principal components* [29]. Its key idea is to project the initial data on a new orthogonal subspace to find the linear combinations that define new summarizing variables, which concentrate the largest possible variance of the original ones.

Therefore, we find that the PCA represents a good solution to assist in solving the limitations caused by the important number of inputs and the measures correlation. We intend to follow this trail as a backstage preprocessing step that would ensure the generation of new reduced training sets that preserves the measure variability.

On the other hand, OLAP measures have multiple linear variations following the different axes of the data cube. Considering a single measure variation may make the prediction process fall into the pitfall of promoting a particular set of dimensions at the expense of the other ones. This could generate a prediction model that may not reflect the complete multidimensional context.

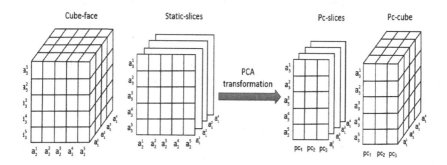

Fig. 5. Overview of the preprocessing stage of a single static-cube

To tackle this issue, we introduce the Neural Approach for Prediction over High-dimensional Cubes (NAP-HC). The NAP-HC's main goal is to overcome the challenges of the application of Neural Networks (NN)s within the context of OLAP cubes. To do so, we design it over the MCEF, which is dedicated to assist the application of classical data mining techniques on data cubes.

The NAP-HC combines the modular aspects of MCEF and MNN to provide a prediction solution that enables the application of NNs on a data cube, while covering all its data presentations. As shown in Fig. 4, NAP-HC is carried out in two major stages. The first one is a preprocessing stage, which is divided, in its turn, into two steps. The first one consists in extracting the MCEF substructures. As for the second step, it consists at applying the PCA on the MCEF substructures to transform their correlated attributes into reduced sets of decorrelated principal components. The second stage is a prediction one, which considers each reduced dataset, obtained over the first stage, as the learning set of an independent NNs module. Then, it trains a NNs combiner system, which considers the outputs of each module as its own inputs, to come out with a unique predicted measure of each targeted cell.

To sum up, the NAP-HC overarching goals are as follows:

1. Generating reduced, information preserving training sets from the original data cube;
2. Adapting NNs to the multidimensional structure of data cubes;
3. Predicting explicit values of non-existent measures;
4. Assessing predicted measures with quality indicators.

4.2   Preprocessing Stage

The main goal of this stage is to generate concentrated, independent, information-preserving data subsets, which can be exploited later as the training sets of the independent modules. As illustrated in Fig. 4, it is based on two main steps. The first one consists in decomposing the complex multidimensional data cube domain into a set of linear sub-domains, defined by the MCEF substructures. As for the second step, it is a dimensions reduction step, which consists in applying PCA on the obtained MCEF substructures.

The NAP-HC exploits MCEF as a modular principled cube explorations technique. First, the dimensions are distributed over three mutually exclusive sets, following the analysis's goals. Each of these sets plays a different role as one of MCEF dimensions classes. Then, all the possible MCEF classes' combinations are considered to define and extract potential static-cubes, which consist of distinct data presentation following the dimensions' distribution over its axes.

The second step of the preprocessing stage is illustrated in Fig. 5. It is a dimensions reduction and data transformation step. Its main goal is to reduce the attributes of each static-cube Variables dimensions and to transform its members into a reduced, concentrated set of principal components. It starts by extracting sequentially each static-cube static-slices by sequentially applying MDX queries.

Static-slices are not dynamic such as classic OLAP slices, so they can be directly considered as disjunctive tables with Variables dimensions as attributes' dimension and Rows dimensions as instances' dimension. The PCA is then applied sequentially on the static-slices, to generate a new type of slices, which we refer to as *pc-slice*. Each pc-slice shares the same Rows and Contexts dimensions' sets with its associated static-slice. However, its Variables dimensions are replaced with a new dimension, referred to as *pc-dimension*. It has the set of retained principal components as attributes. As for the obtained factorial coordinates, they are stored as the values of the pc-slices.

The set of pc-slices generated over the static-slices of the same static-cube, are gathered as a new multidimensional structure that we call *pc-cube*. Actually, a pc-cube is a static structure associated to one particular static-cube. Unlike regular OLAP cubes, pc-cubes are not dynamic and do not support OLAP operations. Their role consists in providing an organized storage solution for the obtained factorial coordinates to track their membership to the original cube cells. They are trackable from the data cube through a new type of measure, which we call *pc-measure*. It is an indexation measure that links each cell to its adequate factorial coordinates in the pc-cubes. It is embedded within the original cells' measures after the application of PCA on each static-slice.

The usages of these new PCA oriented concepts provide an efficient storage solution. It enables discarding each static-cube from the main memory, as soon as its associated pc-cube is generated. The storage of pc-cubes is less expensive than the storage of static-cubes, since they represent their reduced version. Doing so, the preprocessing stage provides reduced, decorrelated predictors that require a minimum storage cost.

Algorithm 1. Static-cube generation and reduction

    Input: Ω^c, Ω^v, Ω^r
    Output: The pc-cube Pcc
1 $C \leftarrow generate\_cube\text{-}face(\Omega^c, \Omega^v, \Omega^r)$;
2 $Pcc \leftarrow \emptyset$;
3 $i = 0$;
4 **foreach** $nonempty$ Θ_i^c of Ω^v **do**
5       $S_i \leftarrow generate\_slice(\Theta_i^c, \Omega^v, \Omega^r)$;
6       $Pcs_i \leftarrow PCA(S_i)$;
7       $Pcc \leftarrow Pcc + Pcs_i$;
8       $i \leftarrow i + 1$;
9 $return(Pcc)$;

The static-cube generation and reduction is provided in Algorithm 1. It requires three inter-dimensional hierarchical predicates Ω^c, Ω^v, Ω^r translating the three MCEF classes as inputs and processes as follows:

- The static-cube C is generated according to the inter-dimensional hierarchical predicates Ω^c, Ω^v and Ω^r.
- Each inter-dimensional hierarchical predicate $\Omega_i^c \subset \Omega^c$ is instantiated to the next nonempty inter-dimensional predicate Θ_i^c.
- The static-slice $S[\Theta_i^c, \Omega^v, \Omega^r]$; S_i is then generated.
- PCA is applied on S_i and the obtained factorial coordinates are stored into the pc-slice pcs_i.
- pcs_i is added to the pc-cube Pcc.
- the output of this algorithm is a fully indexed pc-cube, representing the reduced version of the treated static-cube.

We admit that, similarly to of the conventional OLAP preprocessing phases, this preprocessing stage is a time-consuming one. Therefore, we believe that it should be executed in backstage on a regular basis by the end of each periodic data loading of the data warehouse.

4.3   Prediction Stage

The main goal of this stage is to learn from the outputs of the preprocessing stage, which are the pc-cubes, to come out with unique explicit value for each targeted measure. To the best of our knowledge, PCA has not been yet exploited with MNNs by any previous work.

By virtue of their operation simplicity, their excellent generalization capacity and their ability to approximate any universal function, Multilayer Perceptrons (MLP)s represent one of the popular NNs [30]. Thus, for all the sub-networks that compose our system, we adopt the MLPs architecture. In addition, several theoretical and empirical studies show that a single hidden layer is sufficient to achieve a satisfactory approximation of any nonlinear function [30]. Thus, we associate a three layers MLPs architecture, including a single hidden layer for each sub-network. We also use the gradient back-propagation algorithm [31], that has proven its usefulness in several applications [30,32]. We associate it with the conjugate gradient learning method and the sigmoid activation function.

The prediction system is composed of an interconnection of a set of module-networks and a single combiner-network. The number of module-networks is equal to that of pc-cubes obtained of the preprocessing stage. Each module-network is trained independently. It considers the factorial coordinates as inputs and targets the measure' values. In addition, each module-network has three layers:

1. An input layer, which contains a number of neurons equal to that of of the principal components of the pc-cube associated to the module;
2. A hidden layer, which contains an empirically selected number of neurons;
3. An output layer that contains a single output.

As for the combiner-network, it follows the same architecture as the modules except that its input layer neurons' number is equal to the number of module-networks. It brings together all the module-networks output as its own inputs. Thus, the input vector of the combiner-network is obtained by propagating the factorial coordinates associated to the same cell into all the module-networks. The measure's value of this cell represents the output of the combiner-network. This process is repeated until the combiner-network reaches the convergence status at its turn.

The pseudo-code of the training algorithm is described in Algorithm 2. As inputs, it requires the data cube $\mathcal{C}$, the set of the obtained over the preprocessing stage pc-cubes $\{Pcc\}$ and the Root Mean Squared Error(RMSE) minimum value $RMSE\text{-}min$. For each module-network, NAP-HC starts by selecting a random set of cells as training set from the data cube, $A[]$, and the pc-cube, Pcc, associated to the treated module. For each training cell, the algorithm accesses the pc-measure, pc, and fetches it to get its appropriate factorial coordinates vector, $fc[]$, from the pc-cube. $fc[]$ is then injected into the input layer of the module-network, while targeting the initial measure's value. This process is repeated for each module-network until there are no more training instances or until the RMSE reaches $RMSE\text{-}min$ value. After performing the sequential independent training of all the module-networks, the combiner-network becomes ready to be initialized and trained.

We stress that our approach is not a cube completion technique, i.e. it is not designed to fill all empty measures of a data cube. However, the main goal of

---

Algorithm 2. Training the prediction system

Input: $\mathcal{C}, \{Pcc\}, RMSE - min$

Output: Trained prediction system

1 **foreach** *module* **do**
2     $Pcc \leftarrow select\_ Pcc(\{Pcc\});$
3     $module \leftarrow initialize(module);$
4     $A[] \leftarrow generate\_ random - cells(\mathcal{C});$
5   **while** $((A[] \neq \emptyset)$ **and** $(RMSE(module) < RMSE - min))$ **do**
6        $m \leftarrow get\_ measure(\mathcal{C}, A[]);$
7        $pc \leftarrow get\_ pc - measure(Pcc, A[]);$
8        $fc[] \leftarrow get\_ factorial - coordiantes(Pcc, pc);$
9        $propagate(module, fc, m);$
10        $back - propagate(module, fc, m);$
11        $adjust(module);$

12 $combiner \leftarrow initialize(combiner);$
13 $A[] \leftarrow generate\_ random - cells(\mathcal{C});$
14 **while** $((RMSE(combiner) < RMSE - min)$ **and** $(A[] \neq \emptyset))$ **do**
15     $combiner - input[] \leftarrow \emptyset;$
16   **foreach** *module* **do**
17        $m \leftarrow get\_ measure(\mathcal{C}, A[]);$
18        $pc \leftarrow get\_ pc - measure(Pcc, A[]);$
19        $fc[] \leftarrow get\_ factorial - coordiantes(Pcc, pc);$
20        $combiner - input[] \leftarrow combiner - input[] + propagate(module, fc, m);$
21     $propagate(combiner, combiner - input[], m);$
22     $back - propagate(combiner, combiner - input[], m);$
23     $adjust(combiner);$

24 $return(Trained\ prediction\ system);$

Table 2. Static-cubes description

Static-cube	Contexts	Variables	Rows	# retained components
C_1	Location	Education	Origin	3
C_2	Location	Origin	Education	4
C_3	Education	Location	Origin	10
C_4	Education	Origin	Location	4
C_5	Origin	Location	Education	12
C_6	Origin	Education	Location	4

NAP-HC is to promptly come-out with a predicted value of any empty measure upon the request of the BI analyst.

5   Experimentation

We implemented an experimental prototype of our approach, in Java, on a running on Microsoft Windows 7 with Intel Core 2 Duo, 2 GHz of CPU processor, 4 GB main memory workstation. We used *Microsoft SQL Server Analysis Services 2008*(SSAS) as an OLAP server. We performed our experiments on the database *American Community Surveys 2000–2003*[1], after adapting it to the OLAP context. It is a real-life database of the U.S.A census that concerns the population samples treated between 2000 and 2003.

5.1   Analysis Context

We consider a four dimensions data cube; Location, Origin, Education and Time, with 3.8 million facts. The Location dimension contains the geographic data of the census. Origin dimension contains information about the racial structure of the U.S.A population. Education dimension contains information on the education levels reached by the subjects of the census. We aim at predicting the number of people of a certain race, according to their cities and their levels of education in 2003.

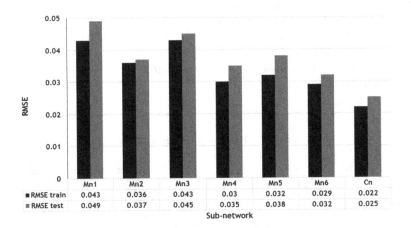

	Mn1	Mn2	Mn3	Mn4	Mn5	Mn6	Cn
■ RMSE train	0.043	0.036	0.043	0.03	0.032	0.029	0.022
■ RMSE test	0.049	0.037	0.045	0.035	0.038	0.032	0.025

Sub-network

Fig. 6. Prediction quality

To be able to analyze and compare the different static-cubes outcomes, we limited each MCEF class to one dimension Location, Education and Origin and we selected the member 2003 of Time dimension. This led to the generation of six static-cubes as summarized by Table 2. We selected the hierarchical levels *Location*, *Education* and *Origin*, respectively. These levels include 51, 14 and 10 members, respectively. We investigated the measure person-count.

---

[1] American Community Surveys is accessible from the official site IPUMS-USA (Integrated Public Use Microdata Series); http://sda.berkeley.edu.

We elaborated a predictive system that faithfully represents our proposed architecture. After the application of the preprocessing stage, we ended up with the 6 pc-cubes from which we retained different numbers of principal components described in Table 2. As for the prediction stage, we have set the number of hidden neurons of each sub-network's hidden layer to the half of its inputs. We used the 10-fold cross-validation technique and the Root Mean Squared Error (RMSE) as a quality indicator. For accuracy reasons, more specifically, to avoid the impact of the random weights initialization of NNs, we ran all the experiments five times and provided the resulting means of RMSE and execution time in this section.

5.2 Prediction Quality

Figure 6 illustrates the prediction performances for all the sub-networks that compose our predictive system. We notice that RMSE values vary remarkably from of a module-network to another one. This is justified by the particularity of the different data structures of each pc-cube. We find that the two module-networks that provide the largest RMSEs, and thus the worst prediction quality, are Mn_1 and Mn_3. We note that these two module-networks consider Origin as their Rows dimension.

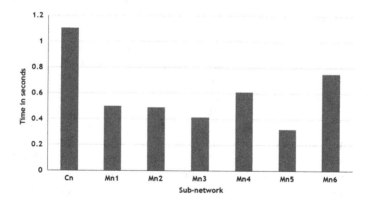

Fig. 7. Training time consumption

We recall that in our proposal, the number of available instances in a training set for a module-network is defined by the number of the Rows dimensions' members. In our case, Origin dimension is the poorer dimension in terms of members' number (10 members). Subsequently, the module-networks that consider it as their Rows dimension have the smallest number of training instances. The poor prediction quality can be due to that this number has been not sufficient to ensure the module-network learning. Inversely, we found that Mn_4 and Mn_6, which consider Location as their Rows dimension, produce the smallest RMSE values among all module-networks.

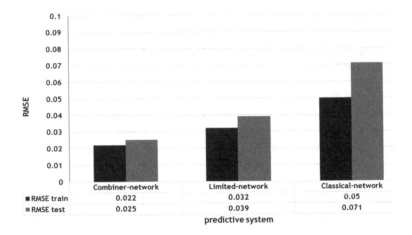

Fig. 8. Performances of the different prediction systems

Interestingly enough, we find that the obtained RMSEs values are generally acceptable. Still, the best prediction performance is achieved by the combiner-network. It surpasses all the module-networks in the training and the test phases. This confirms the efficiency of the modular architecture in generating better prediction by combining the knowledge of all the module-networks. Since, each module-network has become an expert in its particular cube perspective, joining the modules into an ensemble make them compensate each other's limitations through the combiner-network, which combines all static-cubes bi-dimensional knowledge and convert it into a multidimensional one.

As shown on Fig. 7, the training time varies from a sub-network to another one. The most time requiring module-networks are Mn_3 and Mn_5. This is due to the fact that they consider Location, which is the richest dimension in terms of members' number, as Variables dimension, which led to retaining a larger number of principal components after PCA application. Consequently, these module-networks have the largest number of input and hidden neurons among all module-networks, what makes them require a larger number of pc-measures accesses to fetch the factorial coordinates. Moreover, their important number of neurons leads to more complex computations, and thus they consume more time to converge.

Furthermore, we find that the combiner-network is the most time consuming among all sub-networks. This is explained by the fact that at each turn of its training phase, it has to access to all the module-networks' principal components to obtain its own input vector.

5.3   Novel Architecture Contributions

In the previous experiment, we found that several module-networks provide very modest performances compared to the other ones. Following these results, a logical question arises: How would the system perform if we eliminate the module-networks that produce the worst results from the analysis? To answer this question, we designed and trained an additional prediction system, which we call `Limited-network`. The latter is similar to NAP-HC, except that it does not involve Mn_1 nor Mn_3, i.e. the module-networks that performed the least efficient predictions through the previous experiment. Surprisingly enough, as shown in Fig. 8, `Limited-network` provided worst performances than `Combiner-network`. This can be explained by the fact that even if the eliminated module-networks are not useful to perform the prediction task individually, they have a positive role in enhancing the combiner-network knowledge about the multidimensional structure.

To further investigate the contributions of the modular architecture, we trained another system that follows the classical MLPs architecture, which we refer to as `Classical-network`. It considers all the factorial coordinates indexed by the pc-measure of a particular cell as the inputs of one single large MLP. In other words, it merges all the training subsets into a unique large one. As shown in Fig. 8, `Classical-network` provided the worst performances among all the studied architectures. The naïve fusion of the training sets caused the loss of the particularity of the information obtained over each static-cube. Moreover, the large number of inputs limited the MLP generalization abilities. This confirms the positive contributions of MCEF and the efficiency of the combination of MCEF and MNNs.

6   Conclusion and Perspectives

In this paper, we encouraged the exploitation of machine learning techniques to extend OLAP to advanced abilities. The key idea of our proposal is that, these sophisticated techniques can be exploited, in the context of OLAP cubes, even with the challenges raised by their important dimensionality and volumetry. First, we proposed a generalized cube exploration framework, designed to assist the application of machine learning algorithms on OLAP cubes. Then, we exploited it to propose a novel MNNs solution called NAP-HC, which predicts non-existent measures' values over OLAP cubes.

NAP-HC makes use of enhanced procedure to solve the constraints raised by applying sensitive techniques as NNs on a complex data structure like OLAP cubes. It relies on two main stages. A preprocessing one that exploits MCEF to explore the data cube in a principled way, and generate reduced information preserving training subsets by applying PCA on the MCEF substructures. As for the second stage, it exploits the outputs of the first one to train a MNN.

The experimental study proved the efficiency of MCEF in enabling the application of classical data mining algorithms on OLAP cubes. Further, it demonstrated the good prediction performances of NAP-HC and helped to get further insights of the different results obtained over the sub-systems, which form the global model. Furthermore, it compared our proposal against the classical MLPs architecture and confirmed the successful combination between MCEF and MNN.

In future work, we plan to include a framework that explains the reasons of non-existent measures occurrences, similarly to that of [33], which is performed on classical bi-dimensional data. Cuzzocrea and Mansmann state that multidimensional visualization tools provide more comprehensive analysis for multiple cube mining tasks, including discovering new knowledge from large volumes of multidimensional data [19]. Therefore, we intend to equip the NAP-HC with a visualization tool to assist the prediction phase. We also would like to involve the hierarchical structure of data cubes in our system. This way, we could exploit the different levels of aggregation to predict lower/higher-levels facts. Finally, we believe that modeling a theoretical relation between the reduction and the prediction stages could be very useful to optimize our proposal.

References

1. Inmon, W.H.: Building the Data Warehouse. QED Information Sciences Inc, Wellesley, MA, USA (1992)
2. Goil, S., Choudhary, A.: High performance multidimensional analysis and data mining. In: Proceedings of the High Performance Networking and Computing Conference (SC'1998), Orlando, Florida, US, November 1998
3. Sarawagi, S., Agrawal, R., Megiddo, N.: Discovery-driven exploration of OLAP data cubes. In: Schek, H.-J., Saltor, F., Ramos, I., Alonso, G. (eds.) EDBT 1998. LNCS, vol. 1377, p. 168. Springer, Heidelberg (1998)
4. Ben Messaoud, R., Loudcher-Rabaseda, S.: Olemar: An on-line environment for mining association rules in multidimensional data. In: Advances in Data Warehousing and Mining, vol. 2. Idea Group Publishing (2007)
5. Codd, E.F., Codd, S.B., Salley, C.T.: Providing OLAP (on-line Analytical Processing) to User-analysts: An IT Mandate. Codd and Date, Inc., Manchester (1993)
6. Bodin-Niemczuk, A., Ben Messaoud, R., Rabaséda, S.L., Boussaid, O.: Vers l'intégration de la prédiction dans les cubes OLAP. In: EGC. (2008) 203–204
7. Agarwal, D., Chen, B.C.: Latent OLAP: Data cubes over latent variables. In: Proceedings of the 2011 ACM SIGMOD International Conference on Management of Data. SIGMOD 2011, pp. 877–888. ACM, New York (2011)
8. Palpanas, T., Koudas, N., Mendelzon, A.: Using datacube aggregates for approximate querying and deviation detection. IEEE Trans. Knowl. Data Eng. **17**, 1465–1477 (2005)
9. Chen, B.C., Chen, L., Lin, Y., Ramakrishnan, R.: Prediction Cubes. In: Proceedings of the 31st International Conference on Very large Data Bases. VLDB 2005, pp. 982–993 (2005)
10. Cuzzocrea, A.: Providing probabilistically-bounded approximate answers to non-holistic aggregate range queries in OLAP. In: Proceedings of the 8th ACM International Workshop on Data Warehousing and OLAP. DOLAP 2005, pp. 97–106. ACM, New York (2005)
11. Chen, Y., Dong, G., Han, J., Pei, J., Wah, B.W., Wang, J.: Regression cubes with lossless compression and aggregation. IEEE Trans. Knowl. Data Eng. **18**, 1585–1599 (2006)
12. Cuzzocrea, A., Saccà, D.: Balancing accuracy and privacy of OLAP aggregations on data cubes. In: Proceedings of the ACM 13th International Workshop on Data Warehousing and OLAP. DOLAP 2010, pp. 93–98. ACM, New York (2010)
13. Messaoud, R.B., Rabaséda, S.L., Boussaid, O., Missaoui, R.: Enhanced Mining of Association Rules from Data Cubes. In: Proceedings of the 9th ACM International Workshop on Data Warehousing and OLAP (DOLAP'2006), pp. 11–18. ACM Press, Arlington (November 2006)

14. Cuzzocrea, A.: Overcoming limitations of approximate query answering in OLAP. In: 9th International Database Engineering and Application Symposium. IDEAS 2005, pp. 200–209 (July 2005)
15. Jain, A.K.: Fundamentals of Digital Image Processing. Prentice-Hall Inc, Upper Saddle River, NJ, USA (1989)
16. Abdelbaki, W., Ben Messaoud, R., Ben Yahia, S.: A neural-based approach for extending OLAP to prediction. In: Cuzzocrea, A., Dayal, U. (eds.) DaWaK 2012. LNCS, vol. 7448, pp. 117–129. Springer, Heidelberg (2012)
17. Abdelbaki, W., Ben Yahia, S., Ben Messaoud, R.: NAP-SC: a neural approach for prediction over sparse cubes. In: Zhou, S., Zhang, S., Karypis, G. (eds.) ADMA 2012. LNCS, vol. 7713, pp. 340–352. Springer, Heidelberg (2012)
18. Witten, I., Frank, E.: Data Mining: Practical Machine Learning Tools and Techniques, 2nd edn. Elsevier, Morgan Kaufmann, Burlington (2005)
19. Cuzzocrea, A., Mansmann, S.: OLAP visualization: models, issues, and techniques. In: Wang, J. (ed.) Encyclopedia of Data Warehousing and Mining, 2nd edn, pp. 1439–1446. IGI Global, Hershey, PA (2009)
20. Ramakrishnan, R., Chen, B.C.: Exploratory mining in cube space. Data Min. Knowl. Disc. 15(1), 29–54 (2007)
21. Azam, F.: Biologically inspired modular neural networks. Ph.D. thesis, Virginia Polytechnic Institute and State University, Virginia, USA (2000)
22. Bishop, C.: Neural Networks For Pattern Recognition. Oxford University Press, Oxford (1995)
23. Sharkey, A.J. (ed.): Combining Artificial Neural Nets: Ensemble and Modular Multi-Net Systems, 1st edn. Springer-Verlag New York Inc, Secaucus, NJ, USA (1999)
24. Gallinari, P.: The Handbook of Brain Theory and Neural Networks. MIT Press, Cambridge, MA, USA (1998)
25. Melin, P., Castillo, O.: Modular neural networks. In: Hybrid Intelligent Systems for Pattern Recognition Using Soft Computing. Studies in Fuzziness and Soft Computing, vol. 172, pp. 109–129. Springer, Heidelberg (2005)
26. Happel, B.L., Murre, J.M.J.: The design and evolution of modular neural network architectures. Neural Netw. 7, 985–1004 (1994)
27. Tshilidzi, M.: Computational Intelligence for Missing Data Imputation, Estimation, and Management: Knowledge Optimization Techniques. IGI Publishing, Hershey, PA (2009)
28. Wang, Z., Xu, J., Lu, F., Zhang, Y.: Using the method combining PCA with BP neural network to predict water demand for urban development. In: Proceedings of the 2009 Fifth International Conference on Natural Computation. ICNC 2009, pp. 621–625. IEEE Computer Society, Washington (2009)
29. Hotelling, H.: Analysis of a complex of statistical variables into principal components. J. Educ. Psychol. 24(7), 498–520 (1933)
30. Hornik, K., Stinchcombe, M., White, H.: Multilayer feedforward networks are universal approximators. Neural Netw. 2(5), 359–366 (1989)
31. Rumelhart, D., McClelland, J.: Parallel Distributed Processing: Explorations in the Microstructure of Cognition. Foundations. Computational Models of Cognition and Perception. MIT Press, Cambridge (1986)
32. Haykin, S.: Neural Networks: a Comprehensive Foundation. Prentice Hall, Prentice Hall International Editions Series (1999)
33. Ben Othman, L., Ben Yahia, S.: Yet another approach for completing missing values. In: Yahia, S.B., Nguifo, E.M., Belohlavek, R. (eds.) CLA 2006. LNCS (LNAI), vol. 4923, pp. 155–169. Springer, Heidelberg (2008)

Cut-and-Rewind: Extending Query Engine for Continuous Stream Analytics

Qiming Chen[✉] and Meichun Hsu

HP Labs, Hewlett Packard Co., Palo Alto, CA, USA
{qiming.chen,meichun.hsu}@hp.com

Abstract. Combining data warehousing and stream processing technologies has great potential in offering low-latency data-intensive analytics. Unfortunately, such convergence has not been properly addressed so far. The current generation of stream processing systems is in general built separately from the data warehouse and query engine, which can cause significant overhead in data access and data movement, and is unable to take advantage of the functionalities already offered by the existing data warehouse systems.

In this work we tackle some hard problems in integrating stream analytics capability into the existing query engine. We define an extended SQL query model that unifies queries over both static relations and dynamic streaming data, and develop techniques to extend query engines to support the unified model. We propose the *cut-and-rewind* query execution model to allow a query with full SQL expressive power to be applied to stream data by converting the latter into a sequence of "chunks", and executing the query over each chunk sequentially, but without shutting the query instance down between chunks for continuously maintaining the application context across the execution cycles as required by sliding-window operators. We also propose the *cycle-based transaction model* to support Continuous Querying with Continuous Persisting (CQCP) with cycle-based isolation and visibility.

We have prototyped our approach by extending the PostgreSQL. This work has resulted in a new kind of tightly integrated, highly efficient system with the advanced stream processing capability as well as the full DBMS functionality. We demonstrate the system with the popular Linear Road benchmark, and report the performance. By leveraging the matured code base of a query engine to the maximal extent, we can significantly reduce the engineering investment needed for developing the streaming technology. Providing this capability on proprietary parallel analytics engine is work in progress.

1 Introduction

Streaming analytics is a data-intensive computation chain from event streams to analysis results. In response to the rapidly growing data volume and the pressing need for lower latency, Data Stream Management Systems (DSMSs) provide a paradigm shift from the load-first analyze–later mode of data warehousing [8, 16, 17, 19].

1.1 The Problem

However, the current generation of DSMS is in general built separately from the data warehouse query engine, due to the difference in handling stream data and static data;

© Springer-Verlag Berlin Heidelberg 2015
A. Hameurlain et al. (Eds.): TLDKS XXI, LNCS 9260, pp. 94–114, 2015.
DOI: 10.1007/978-3-662-47804-2_5

as a result, the data transfer overhead between the two has become a performance and scalability bottleneck [4, 6, 10]. The standalone DSMS's also lack the full SQL expressive power and DBMS functionalities of managing persistent data. It does not have the appropriate transaction support for continuously persisting and sharing results along with continuous querying. As stream processing evolves from simple to complex, these functionalities are likely to be redeveloped.

In this paper we tackle the following technical challenges in integrating stream processing with data warehouse query engine:

- A query engine manages relations (tables) which contain well defined sets. However, a stream is unbounded, and never reaches the "end of data", which would pose problems with the existing query model and transaction model.
- Stream processing is often based on windows, and there is a need to apply a query repeatedly to chunks of unbounded stream data that fall in consecutive windows. Stream analytics also requires operators that are history sensitive, such as sliding window operators, and there is a need to continuously and efficiently maintain the state or a synopsis of the data that falls in the previous windows.
- During stream processing, there is a need to persist periodically to allow the analysis results to be visible to other concurrent applications, sometimes even to another branch of the same query. This will require extended transaction semantics that is not supported with existing query engines.

1.2   State of the Art

Since a stream query is defined on unbounded data and in general limited to non-transactional event processing, the current generation of DSMSs is mostly built from scratch independently of the database engine. Big players along this direction include System S (IBM) [15], STREAM (Stanford) [3], TelegraphCQ (Berkeley) [5], as well as Aurora, Borealis, etc. [1, 2, 7, 11, 17]. Two recently reported systems, the TruSQL engine [16] developed by Truviso Inc, USA, and the DataCell engine [19] developed by CWI, Netherlands, do leverage database technology but are characterized by providing a workflow like service for launching a SQL query for each chunk of the stream data during stream processing. To the best of our knowledge, none of the existing approaches has leveraged the query engine without introducing an additional loosely-coupled "middleware" layer. Oracle currently offers a "continued query" feature but it is based on automatic view updates and is not the same feature as stream processing.

Managing data-intensive stream processing outside of the query engine causes the data copying and moving overhead, and fails to leverage the full SQL and DBMS functionality.

Processing streams by multiple queries may incur performance penalty due to the overhead for frequent query setup and teardown, and more seriously, cause the semantic difficulty in chunk-wise data manipulation. Since the backend query execution processes are in isolated memory contexts, processing each data chunk by an individual query instance cannot maintain the application context, e.g. the data buffered with User Defined Functions (UDFs) continuously across multiple query instances, thus unable to deal with sliding-window like operations.

To the best of our knowledge, none of the existing approaches has solved the difficulty of processing stream in terms of truly continued SQL query with chunk-wise semantics but continuously tracked application context, by leveraging the query engine without introducing an additional loosely-coupled "middleware" layer.

1.3   The Solution

We view a query engine essentially as a streaming engine, although this potential has not been thoroughly explored. With this vision, we advocate an extended SQL model that unifies queries over both streaming and static relational data, and a new architecture for integrating stream processing and DBMS to support continuous, "just-in-time" analytics with window-based operators and transaction semantics.

Our proposed stream model is based on dividing an infinite stream of relation tuples with a criterion, e.g. by every 1-minute time window, into an unbounded sequence of chunks. The semantics of applying the query to the unbounded stream lies in applying the query to those infinite chunks which continuously generates an unbounded sequence of query results, one on each *chunk* of the stream data.

Our goal is to support the above semantics using a continuous query that runs cycle by cycle for processing the stream data chunks, each data chunk to be processed in each cycle, in a single, long-standing query instance. In this sense we also refer to the *data chunking criterion* C as the *query cycle specification*. The cycle specification can be based on time or a number of tuples, which can amount to as small as a single tuple, and as large as billions of tuples per cycle. The stream query may be terminated based on specification in the query (e.g. run for 300 cycles), user intervention, or a special end-of-stream signal received from the stream source.

Specifically, our solutions include the following.

- We start with providing unbounded relation data to feed queries continuously. The first step is to integrate the notions of stream data source, and use function-scan instead of table-scan, for turning captured events into unbounded sequence of relation tuples to feed to stream queries without first storing them on disk.
- We develop UDF shells [9] to deliver operators with stream semantics (e.g. moving average, notification) that are not available in conventional SQL. We allow a UDF to cache the state in the application context for carrying out history-sensitive operations, such as sliding window oriented operations, along the stream processing pipeline. We also allow a UDF to emit the current or accumulated computation results continuously on the per-tuple basis - once a tuple from the stream has been received and/or processed.
- We propose the *cut-and-rewind* query model, namely, cutting a query execution based on some granule ("chunk") of the stream data (e.g. in a time window), and then rewinding the state of the query without shutting it down, for processing the next chunk of stream data. This mechanism, on one hand, allows applying a query continuously to the stream data chunks falling in consecutive time windows, within a single, long-standing query; on the other hand, allows retaining the application context (e.g. data buffered with UDFs) continuously across the execution cycles to perform sliding-window oriented, history sensitive operations.

– To support *Continuous Querying with Continuous Persisting* (CQCP), we introduce the cycle-based transaction model with the *cycle-based isolation* mechanism, which makes the heap-inserted, chunk-wise database updates accessible by other applications as soon as the corresponding cycle execution commits. Note however that a continuous query may emit non-transactional messages or events to external receivers before "commit" – such messages are not bound by transaction semantics.

A significant advantage of the unified model lies in that it allows us to exploit the full SQL expressive power on each data chunk. The output is also a stream consisting of a sequence of chunks, with each chunk representing the query result of one execution cycle. While there may be different ways to implement our proposed unified model, our approach is to generalize the SQL engine to include support for stream sources. The approach enables queries over both static and streaming data, retains the full SQL power, while executing stream queries efficiently.

The proposed *cut-and-rewind* approach enables us to support truly continuous query in a completely different way from other DSMSs, and seamlessly integrate the stream processing capability into a full–functional database system, creating a powerful and flexible system that can run SQL over tables, streams (tuple by tuple or chunk by chunk), and the combination of the two.

In this paper we have limited a query to refer to a single stream and thus a single cycle specification. In general, our model allows multiple stream queries to refer to the same source, and these queries can interact through database tables which may be memory resident; our model also allows a single query to refer to multiple stream sources with different cut criteria. Various pairing patterns [15] and the corresponding operations to allow multiple streams or hybrid queries to interact have been investigated and are to be reported separately.

We report our experience in leveraging the PostgreSQL engine for supporting stream processing. The proposed *cut-and-rewind* mechanism has been implemented with minimal engine extension, resulting in a tightly integrated, highly efficient platform with the advanced stream processing capability as well as the full DBMS functionality. We demonstrated the merit of our platform using the popular Linear Road benchmark. Providing this capability on a proprietary parallel database engine is currently being explored.

The rest of this paper is organized as follows: Sect. 2 reports our approach in handling stream source and stream analytic functions by extending a DBMS with new source functions and UDFs for stream operations; Sect. 3 proposes the *cut-and-rewind* approach; Sect. 4 deals with the transaction issues in cycle-based stream processing; Sect. 5 shows how the proposed approach is applied to the popular Linear Road stream processing benchmark, and discusses the experiment results; Sect. 6 concludes the paper.

2   Stream Processing as Continuous Querying

A SQL query is parsed and optimized into a query plan that is a tree of operators. The scan operator at the leaf of the tree gets and materializes a block of data to be delivered to the upper layer tuple by tuple. A non-blocking relational operator or a function, e.g. a UDF, is invoked multiple times in a query execution on the per-tuple basis, which forms a dataflow pipeline, and in this sense, similar to stream processing.

However, there exist some fundamental differences between the conventional query processing and the stream processing. First, a query is defined on bounded relations but stream data are unbounded; next, stream processing adopts window-based semantics, i.e. processing the incoming data chunk by chunk falling in consecutive time windows; however, the SQL operators are either based on one tuple (such as filter operators) or the entire relation; Further, stream processing is also required to handle sliding window operations continuously across chunk based data processing; and finally, endless stream analytics results must be continuously accessible along their production, under specific transaction semantics.

Let us use a simplified traffic system example to illustrate our unified query over stored and stream data, where the total amount of toll charged for each highway segment per minute are computed, given a segment toll table and events that report vehicles' entering a segment.

- *C (vid,sid,ts)*, contains the event that a car (*vid*) enters a tolled segment (*sid*) with a timestamp in second (*ts*),
- *T (sid, charge)* contains the highway segment info where *charge* is the toll per car for segment *sid*.

We express the example first as a query over static relations only, and then as a hybrid query that includes a stream source. The graphical representation of the two queries is shown in Fig. 1.

For the first query *Q1* (shown on the left of Fig. 1), the inputs are two stored relations, *C* and *T*. However, if the table *C* above is not a stored relation, but replaced

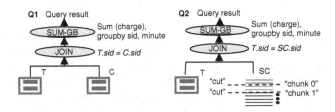

Fig. 1. Querying static table vs. querying data stream chunk by chunk

by a real-time stream source, while *T* remains a stored relation, then the above application becomes a streaming application. The above static SQL query is adapted to a streaming query simply by defining *SC* as a *stream* (instead of a table) with the same schema as *C* and changing the reference to *C* as follows (shown on the right of Fig. 1):

Q1:
SELECT sid, floor(ts/60) AS minute, SUM(charge)
FROM T, C WHERE C.sid = T.sid
GROUP BY sid, minute

Q2:
SELECT sid, floor(ts/60) AS minute, SUM(charge)
FROM T, STREAM (SC, cycle-spec) WHERE SC.sid = T.sid
GROUP BY sid, minute

In the above query, we replace the disk-resided database table by a special kind of table function *STREAM()*, called Stream Source Function (SSF), that listens or reads data/events sequence. Further, *STREAM(SC, cycle-spec)* specifies that the stream source *SC* is to be "cut" into an unbounded sequence of *chunks* SC_{C0}, SC_{C1}, ..., where all tuples in SC_{Ci} occur before any tuple in SC_{Ci+1} in the stream. The "cut point" is specified in the *cycle-spec*. Let *Q1* above be denoted as a query function over table *C*, i.e., *Q1(C)*. The execution semantics of *Q2* is defined as executing *Q1(SC_{Ci})* in sequence for all SC_{Ci}'s in the stream source *SC*.

In general, given a query *Q* over a set of relation tables $T_1,..,T_n$ and an infinite stream of relation tuples *S* with a criterion ϑ for cutting *S* into an unbounded sequence of chunks, e.g. by every 1-minute time window, $< S_0, S_1, ..., S_i, ... >$ where S_i denotes the *i-th* "chunk" of the stream according to the chunking-criterion ϑ. S_i can be interpreted as a relation. The semantics of applying the query *Q* to the unbounded stream *S* plus the bounded relations $T_1,..,T_n$ lies in

$$Q(S, T_1, .., T_n) \rightarrow \ <Q(S_0, T_1, .., T_n), \ ... Q(S_i, T_1, .., T_n), \ ... >$$

which continuously generates an unbounded sequence of query results, one on each *chunk* of the stream data.

2.1 Stream Source Function

For providing unbounded relation data to fuel queries continuously, the first step is to replace the database table, which contains a set of tuples on disk, by the special kind of table function, called Stream Source Function (SSF) that returns a sequence of tuples to feed queries without first storing on disk. A SSF can listen or read data/events sequence and generate stream elements tuple by tuple continuously. A SSF is called multiple, up to infinite, times during the execution of a continuous query, each call returns one tuple. When the end-of-cycle event or condition is seen, the SSF signals the query engine to terminate the current query execution cycle.

We rely on SSF and query engine for continuous querying on the basis that "as far as data do not end, the query does not end", rather than employing an extra scheduler to launch a sequence of one-time query instances. The SSF scan is supported at two levels, the SSF level and the query executor level. A data structure containing function call information, *hFC*, bridges these two levels. *hFC* is initiated by the query executor and passed in/out the SSF for exchanging function invocation related information. We use this mechanism for minimizing the code change, but maximize the extensibility, of the query engine.

2.2 Stream Analytics Through UDF

One important characteristics of stream processing is the use of stream-oriented history-sensitive analytic operators such as moving average or change point detection. While the standard SQL engine contains a number of built-in analytic operators, stream history-sensitive operators are not supported. Using UDFs is the generally accepted mechanism to extend query operators in a DBMS. A UDF can be provided with a data buffer in its function closure, and for caching stream processing state (synopsis).

Furthermore, it is also used to support one or more *emitters* for delivering the analytics results to interested clients in the middle of a cycle, which is critical in satisfying stream applications with low latency requirement.

Stream processing involves operations on (time) windows, including sliding windows, and therefore is history sensitive. This represents a different requirement from the regular query processing that only cares about the current state. We use UDFs to add window operators and other history sensitive operators, buffering required raw data or intermediate results within the UDF closures.

A scalar UDF is called multiple times on the per-tuple basis, following the typical FIRST_CALL, NORMAL_CALL, FINAL_CALL skeleton. The data buffer structures are initiated in the FIRST_CALL and used in each NORMAL_CALL. A window function defined as a scalar UDF incrementally buffers the stream data, and manipulates the buffered data chunk for the required window operation. Since the query instance remains alive, as supported by our *cut-and-rewind* model, the UDF buffer is retained between cycles of execution and the data states are traceable continuously (we see otherwise if the stream query is made of multiple one-time instances, the buffered data cannot be traced continuously across cycle boundaries). As a further optimization, the static data retrieved from the database can be loaded in a window operation initially and then retained in the entire long-standing query, which removes much of the data access cost as seen in the multi-query-instances based stream processing.

We propose to run a SQL query cycle by cycle for deriving a sequence of data-chunk based results, but never shutting down the query instance in order to have the per-tuple based data processing history continuous tractable.

UDFs can be used to develop a library of reusable stream operators and further allow the unified query model to be extended. As will be illustrated in our Linear Road (LR) implementation, the 5-minute moving average speed is provided through a moving average UDF, atop the per-minute average speed, the latter computed using the standard SQL average-groupby function in one query cycle.

3   Cycle Based Continuous Query

To support the cycle based execution of stream queries, we propose the *cut-and-rewind* query execution model, namely, cut a query execution based on the cycle specification (e.g. by time), and then rewind the state of the query without shutting it down, for processing the next chunk of stream data in the next cycle.

Under this *cut-and-rewind* mechanism, a stream query execution is divided into a sequence of *cycles*, each for processing a chunk of data only; it, on one hand, allows applying a SQL query to unbounded stream data chunk by chunk within a single, long-standing query instance; on the other hand, allows the application context (e.g. data buffered within a User Defined Function (UDF)) to be retained continuously across the execution cycles, which is required for supporting sliding-window oriented, history sensitive operations. Bringing these two capabilities together is the key in our approach.

Cut *Cutting* stream data into chunks is originated in the SSF at the bottom of the query tree. Upon detection of end-of-cycle condition, the SSF signals *end-of-data* to the query engine through setting a flag on the function call handle, that, after being interpreted by the query engine, results in the termination of the current query execution cycle.

If the cut condition is detected by testing the newly received stream element, the *end-of-data* event of the current cycle would be captured upon receipt of the first tuple of the next cycle; in this case, that tuple will not be returned by the SSF in the current cycle, but buffered within the SSF and returned as the first tuple of the next cycle. Since the query instance is kept alive, that tuple can be kept across the cycle boundary.

Rewind Upon termination of an execution cycle, the query engine does not shut down the query instance but *rewinds* it for processing the next chunk of stream data. Rewinding a query is a top-down process along the query plan instance tree, with specific treatment on each node type. In general, the intermediate results of the standard SQL operators (associated with the current chunk of data) are discarded but the application context kept in UDFs (e.g. for handling sliding windows) are retained. The query will not be re-parsed, re-planned or re-initiated.

Note that rewinding the query plan instance aims to process the next chunk of data, rather than re-deliver the current query result; therefore it is different from "rewinding a query cursor" for re-delivering the current result set from the beginning. For example, the conventional cursor rewind tends to keep the hash-tables for a hash-join operation but our rewind will have such hash-tables discarded since they were built for the previous, rather than the next, data chunk.

As mentioned above, the proposed *cut-and-rewind* approach has the ability to keep the continuity of the query instance over the entire stream while dividing it to a sequence of execution cycles. This is significant in supporting history sensitive stream analytic operations, as discussed in the previous section.

4 Continuous Querying with Continuous Persisting (CQCP)

One problem of the current generation of DSMSs is that they do not support transactions. Intuitively, as stream data are unbounded and the query for processing these data may never end, the conventional notion of transaction boundary is hard to apply. In fact, transaction notions have not been appropriately defined for stream processing, and the existing DSMSs typically make application specific, informal guarantees of correctness.

However, to allow a hybrid system where stream queries can refer to static data stored in a database, or to allow the stream analysis results (whether intermediate or final) to persist and be visible to other concurrent queries in the system in a timely manner, a transaction model which allows the stream processing to periodically "commit" its results and makes them visible is needed.

Note that if a stream application does not use static data in the database, or does not need to persist results and make them visible to other concurrent applications, then transaction semantics are not needed. In our design, the transaction semantics is used, and thus transaction management overhead is incurred, only when a stream application requires persistent data management.

4.1   Query Cycle Based Transaction Model

Lacking formal transaction semantics is a problem of the current generation of stream processing systems, as they typically make application specific, informal guarantees of correctness.

Conventionally a query is placed in a transaction boundary; the query result and the possible update effect are made visible only after the commitment of the transaction (although weaker transaction semantics do exist). Since the query for processing unbounded stream data may never end, the conventional notion of transaction boundary is hard to apply.

In order to allow the result of a long-running stream query to be incrementally accessible, we introduce the cycle-based transaction model coupled with the *cut-and-rewind* query model, which we call *continuous querying with continuous persisting*. Under this model a stream query is "committed" one cycle at a time in a sequence of "micro-transactions". The transaction boundaries are consistent with the query cycles, thus synchronized with the chunk-wise stream processing. The per-cycle stream processing results are made visible as soon as the cycle ends.

For example, in Q2 above, the query result, which is the total charge per highway segment, is made visible every cycle; if the cycle specification is per minute, then the total charge per segment is made visible per minute, and it can also be persisted at the minute boundary.

4.2   Staging Results Without Data Copy/Move

With the cloud service, the analytics results are accessed by many clients through PCs or smart phones. These results are read-only time series data, stored in the read-sharable tables incrementally visible to users as they become available. Since the analytics results are derived from unbounded stream of events, they are themselves unbounded and thus must be staged step by step along with their production. Very often, only the latest data is "most wanted". For scaling up CaaaS, efficient data staging is the key.

Data staging is a common task of data warehouse management. The general approach is stepwise archival of the older data, which, however, incurs data moving and copying overhead. While this approach is acceptable for handling slowly-updated data in data warehousing, it is not efficient for supporting real-time stream analytics.

To avoid the data moving and copying overhead in data staging, we have developed a specific mechanism characterized by *staging through metadata manipulation without real data movement*. As shown in Fig. 2, we provide a list of tables for keeping the stream analytics results generated in a given number of query execution cycles (e.g. generated in 60 per-minute cycles, i.e. one hour). These tables are arranged as a "table-ring" and used in a *round-robin fashion*. For example, to keep the results for the latest 8 h of notifications, 9 tables say T_1, T_2, ..., T_9, are allocated in a buffer-pool, such that at a time, T_1 stores the results of the current hour, say h, T_2 stores the results of the hour $h-1$, ..., T_8 stores the results of the hour $h-8$, the data in T_9 are beyond the 8-hour range thus being archived asynchronously during the current hour. When the hour changes, the archiving of T_9 has presumably finished and T_9 is reassigned for storing the results of the new, current hour.

The hourly based timestamp of these tables are maintained either in the data dictionary or a specifically provided system table. In the above data staging, only the "label" of a table is switched for representing the time boundary (i.e. the hour) of its content, without moving/copying the content to another table or file thus avoiding the read/write overhead.

Further, a stable SQL interface is provided for both the client-side users and the server-side queries. Assuming the table holding the summarized traffic status in the current hour is named "*current_road_condition*", this name remains the same at all the times but points to different physical tables from time to time. This may be accomplished by associating the table holding the latest results to "*current_road_condition*" through metadata lookup, or by system internal query modification.

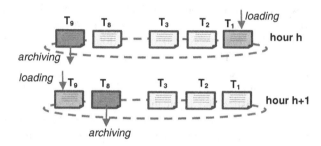

Fig. 2. Table-ring approach for staging analytics results through metadata manipulation without data copy/move

We have extended the query engine to support the above table ring for the client-side query. The continuous query uses the INSERT-INTO clause to capture the query results at each cycle. (See Sect. 2.5 for an example).The "into-relation" is closed prior to a cycle-based transaction commits and it re-opens after the transaction for the next cycle starts. Between the *complete_transaction()* call and the *reopen_into_relation ()* call, the number of execution cycles is checked, and if the specified staging time boundary is reached, the switching of "into-relations", i.e. the query destinations, takes place, where the above data dictionary or specific system table is looked up, and the "next" table ID is obtained and passed to the *reopen_into_relation()*. Thereafter another

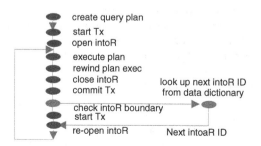

Fig. 3. Cycle-based query execution, transaction, staging

into-relation will act as the query destination. This way, the query runs cycle by cycle to process the input data stream chunk by chunk.

Overall, the cycle-based query execution, transaction commitment and multi-cycle based data staging are illustrated in Fig. 3.

5 Example and Experiments

5.1 Modeling the Linear Road Benchmark

We use the widely-accepted Linear-Road (LR) benchmark [18] to demonstrate our extended query engine. The LR benchmark models the traffic on express ways for the 3-hour duration; each express way has two directions and 100 segments. Cars may enter and exit any segment. The position of each car is read every 30 s and each reading constitutes an event, or stream element, for the system. A car position report has attributes *vid* (vehicle ID), *time* (seconds), *speed* (mph), *xway* (express way), *dir* (direction), *seg* (segment), etc. The benchmark requires computing the traffic statistics for each highway segment, i.e. the number of active cars, their average speed per minute, and the past 5-minute moving average of vehicle speed. Based on these per-minute per-segment statistics, the application computes the tolls to be charged to a vehicle entering a segment any time during the next minute, and notifies the toll in real time (notification is to be sent to a vehicle within 5 s upon entering the segment). The application also includes accident detection; an accident occurring in one segment will impact the toll computation of that segment as well as a few downstream segments. An accident is flagged when multiple cars are found to have stopped in the same location. The graphical representation of our implementation of the LR stream processing requirement is shown in Fig. 4 together with its corresponding stream query.

```
INSERT INTO toll_table SELECT minute, xway, dir, seg, lr_toll(r.traffic_ok, r.cars_volume)
FROM (
    SELECT minute, xway, dir, seg, cars_volume,
            lr_moving_avg(xway, dir, seg, minute, avg_speed) as mv_avg, traffic_ok
    FROM (
        SELECT floor(time/60)::integer AS minute, xway, dir, seg,
                AVG(speed) AS avg_speed, COUNT(distinct Vid)-1) AS cars_volume,
                MIN(trffic_ok) AS traffic_ok
    FROM (
            SELECT xway, dir, seg, time, speed, vid,
                    lr_acc_affected(vid,speed,xway,dir,seg,pos) AS traffic_ok
            FROM STREAM_CYCLE_lr_data(60, 180)
            WHERE lr_notify_toll(vid, xway, dir, seg, time)>=0
        ) s
        GROUP BY minute, xway, dir, seg
    ) p
) r
WHERE r.mv_avg > 0 AND r.mv_avg < 40;
```

This query provides the following major functions.

– **Stream Source Function** - The streaming tuples are generated by the SSF *STREAM_CYCLE_lr_data(time, cycles)*, from the LR data source file with time-stamps, where parameter *"time"* is the time-window size in seconds; *"cycles"* is the number of cycles the query is supposed to run. For example, *STREAM_CY-CLE_lr_data(60, 180)* delivers the position reports one-by-one until it detects the end of a cycle (60 s), and performs a "cut", then onto the next cycle, for a total of 180 cycles (for 3 h).

– **Segment statistics and toll generation** - As illustrated by the left hand side of Fig. 4, the tolls are derived from the segment statistics, i.e. the number of active cars, average speed, and the 5-minute moving average speed, as well as from detected accidents, and dimensioned by express way, direction and segment. We leveraged the *minimum, average* and *count-distinct* aggregate-groupby operators built into the SQL engine, and provided the *moving average* (lr_moving_avg) operator and the *accident detection* (lr_accident) operator in UDFs.

– **Toll persisting** - Required by the LR benchmark, the segment tolls of minute *m* should be generated within 5 s after *m*. The toll of a segment calculated in the past minute is applied to the cars currently entering into that segment. The generated tolls are inserted into a *segment toll table* (SegToll) with the transaction committed per cycle (i.e., per minute). Therefore the tolls generated in the past minutes are visible to the current minute.

– **Toll notification** - As shown on the right side of Fig. 4, the per-car toll notification is provided by the UDF *lr_notify_toll()* appearing in the following phrase

$$\text{WHERE } lr\_notify\_toll(vid, \ xway, \ dir, \ seg, \ time) \ > \ = \ 0$$

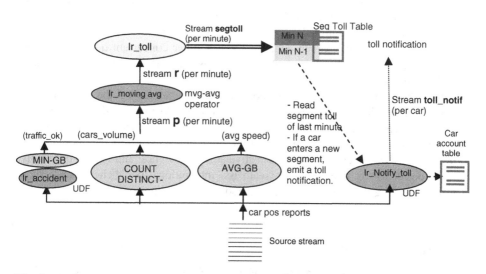

Fig. 4. Cycle based stream query for LR benchmark, for both the generation of per-minute, per cycle tolls common to all cars, and the per car based retrieval of resulting tolls

This UDF keeps enough information about active cars so as to detect the event of a car entering a new segment; and for each car entering a new segment, it emits a toll notification while persisting the toll to a table (carAccount table) for future account balance queries. This UDF reads the segment tolls of the previous minute within the FIRST_CALL part of the UDF (represented by the dash line), enabling it to use the information produced by the previous cycle of the stream query. Since this UDF is a per-tuple UDF (i.e., the NORMAL_CALL part of the UDF is invoked per input tuple), the toll notification is emitted immediately after the position report is received from the source stream, and does not wait for the current cycle (minute) to terminate. This UDF also persists the toll into the car account table. While the toll is notified immediately upon receiving the car position report, persisting the toll is committed once per cycle, in accordance to our CPCQ model.

Multiple features of our cycle-based stream processing approach are illustrated in this query:

- **Cut-and-Rewind.** This query repeatedly applies to each data chunks falling in 1-minute time-window as an execution cycle, and rewinds 180 times in the single query instance; the sub-query with alias p uses the standard SQL aggregate-groupby function to yield the number of active cars and their average speed for every minute dimensioned by segment, direction and express way. The SQL aggregate functions are computed for each cycle with no context carried over from one cycle to the next.
- **Sliding Window Function (per-tuple history sensitive).** The sliding window function $lr\_moving\_avg()$ buffers the up to 5 per-minute average speed for accumulating the dimensioned 5-minute moving average; since the query is only rewound but not shut down, this buffer is retained continuously across query cycles – this is a critical advantage of cut/rewind over shutdown/restart.
- **Continuous Querying with Continuous Persisting.** The top-level construct of the LR query is actually the INSERT-SELECT phrase; with our engine extension, it persists the result stream returned from the SELECT query (r) to the toll table on the per-cycle basis. The transactional LR query commits per cycle to make the cycle based result accessible to subsequent cycles or other concurrent queries after the cycle ends. This cycle-based isolation level is supported with the appropriate locking mechanism.
- **Self-Referencing.** The per-car toll notification is generated by the UDF $lr\_noti\-fy\_toll()$. It efficiently accesses the segment toll in the *last minute* directly from the toll table. This kind of self-referencing provides a handshake mechanism for the *producer* part and the *consumer* part of the same query to rely on the query engine to synchronize, to perform history sensitive stream analytics, and to gain extremely high performance due to their seamless integration. We believe that such self-referencing represents a common paradigm in stream processing.

5.2   Experimental Setup

The experimental results are measured on HP xw8600 with 2 x Intel Xeon E54102 2.33 Ghz CPUs and 4 GB RAM, running Windows XP (x86_32) and PostgreSQL 8.4.

The input data are downloaded from the benchmark's home page. The "L = 1" setting was chosen for our experiment which means that the benchmark consists of 1 express way (with 100 segments in each direction). The event arrival rate ranges from a few per second to peak at about 1,700 events per second towards the end of the 3-hour duration. Figure 6 (Left) shows the distribution of data volume per minute, i.e. the per-minute throughput.

The LR data can be supplied in the following two modes:

- Stress test mode: the data are read by the SSF from a file continuously without following the real-time intervals (continuous input)
- Real-time input: the data are received from a data driver outside of the query engine with real-time intervals. Each car position report carries a system timestamp assigned by the data driver when the event is generated, which could be compared with the system timestamps generated during when toll notification is emitted, for measuring the response time.

We report our experimental results in these 2 different modes.

5.3 Performance Under Stress Test Mode

Time for computing segment tolls. Calculating the segment statistics and tolls has been recognized as the computation bottleneck of the benchmark in the literature. The LR benchmark requires the segment toll to be calculated based on the segment statistics and traffic status (whether affected by accidents) every minute. We took the left-hand-side of our LR model in Fig. 4 and ran that branch of the query up until the toll is computed, under the stress test mode. The total computation time with L = 1 setting is shown in Fig. 5 (Left). It shows that our system is able to generate the per-minute per-segments tolls for the total 3 h of LR data (approx. 12 Million tuples) in a little over 2 min.

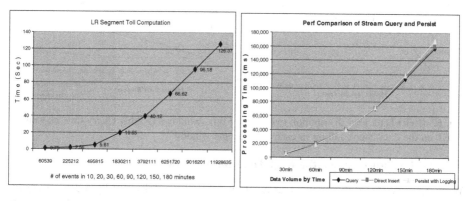

Fig. 5. (Left) Total time of toll computation. **(Right)** Performance comparison of querying-only and query + persisting (with continuous input)

Performance of Persisting with Heap-Insert Unlike other reported DSMSs where the stream processing results are persisted by connecting to a separate database and issuing queries, with the proposed cycle-based CQCP approach, the continuous, minute-cycle based query results are stored through efficient heap-insert.

From Fig. 5 (Right) we can see that persisting the cycle based stream processing results either by inserting with logging (using INSERT INTO with extended support by the query engine) or by direct inserting (using SELECT INTO with extended support by the query engine – not shown in this query), does not add significant performance overhead compared to querying only. This is because we completely push stream processing down to the query engine and handle it in a long running query instance with direct heap operations, with negligible overhead for data movement and for setting up update queries.

Post Cut Elapsed Time. In cycle-based stream processing, the remaining time of query evaluation after the input data chunk is cut, called Post Cut Elapsed Time (PCET), is particularly important since it directly affects the delta time for the results to be accessible after the last tuple of the data chunk in the cycle has been received.

Figure 6 (Left) shows the input data volume over 1-minute time windows (i.e., the stream workload). Figure 6 (Right) shows the query time, as well as the PCET, for processing each 1-minute data chunk. It can be seen that the PCET (the blue line) is well controlled around 0.2 s., meaning that the maximal response time for the segment toll results, as measured from the time a cycle (a minute) ends, is around 0.2 s.

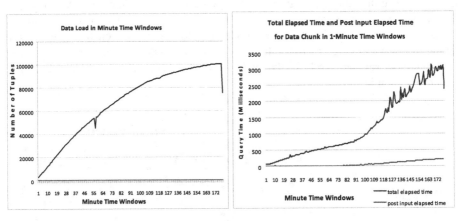

Fig. 6. (**Left**) Data load distribution over minute time windows (**Right**) Query time as well as PCET on the data chunk falling in each minute time window

5.4   Performance Under the Real-Time Input Mode

With real-time input, the events (car position reports) are delivered by a data driver in real-time with additional system-assigned timestamps. The query runs cycle by cycle on each one-minute data chunk. Figure 7 shows the maximal toll notification response time in each of the 180 1-minute windows.

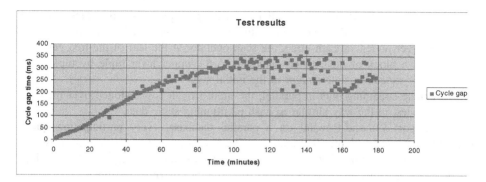

Fig. 7. Maximal toll notification response time in consecutive one-minute time windows

The maximal response time of toll notification really depends on the PCET measure introduced above, i.e. it is essentially the delay after a cycle is "cut" in completing the segment toll part of the query of that cycle. This is because in the beginning of each cycle, the toll notification cannot be emitted until the segment toll generation of the last cycle completes. In the first 2 h the toll notification response time is rather small, and with the increased data load in the last hour, it reaches the maximal value of about 0.3 s, which is still well within the 5-second latency requirement of the benchmark. Note that the maximal notification latency is not the average response time of notification. On the average, the notification response time is near zero, as the ones after the beginning of each cycle are not measurable by millisecond.

The experimental results indicate that our approach is highly competitive to any reported one. This is because we completely pushed stream processing down to the query engine with negligible data movement overhead and with efficient direct heap-insert. We eliminated the middleware layer, as provided by all other systems, for scheduling time-window based querying.

6 Cycle Based Map-Reduce

We rely on the Map-Reduce (MR) computation to scale out CaaaS. With the original MR model; static data are partitioned "horizontally" over cluster nodes for parallel computation; while enhancing the computation bandwidth by divide-and-conquer, it is not defined on unbounded stream data.

We envisage that Cut-and-Rewind (CR) provides a powerful mechanism for MR to reach stream analytics. We have investigated the combination of MR and CR on parallel database platform as well as on network distributed MR infrastructure.

6.1 Cut-Rewind a Parallel Query

A parallel query with UDFs can naturally express Map-Reduce computation. To explain how to apply CR to a parallel query engine for stream processing, let us review the parallel query execution process. A SQL query is parsed and optimized into a query plan that is a tree of operators. In parallel execution multiple sub-plan instances, called fragments, are distributed to the participating query executors and data processors on multiple server nodes; at each node, the scan operator at the leaf of the tree gets and materializes a block of data, to be delivered to the upper layer tuple by tuple. The global query execution state is kept in the initial site.

To handle streaming data in parallel, the input stream is partitioned over multiple machine nodes, in the way similar to hash partitioning static data.

To support Cut-and-Rewind on a parallel database, every participating query engine is facilitated with the CR capability. The same *cut* condition is defined on all the partitioned streams. Note that if the cycle based continuous querying is "cut" on time window, the stream cannot be partitioned by time, but by other attributes.

A query execution cycle ends after *end-of-cycle* is signaled from *all* data sources, i.e. all the partitioned streams are "cut". As the *cut* condition is the same across all the partitioned streams, the cycle-based query executions over all nodes are well synchronized through data driven.

To parallelize the LR stream analysis, we hash partition the data stream by vehicle-id (vid); use the Map function to compute and pre-aggregate the segment traffic statistics per minute (without accident detection); use the Reduce function to globally aggregate the segment statistics, group by express-way, direction and segment, then calculate per segment moving average speed and finally the toll. The whole map-reduce implementation of the application is expressed in a *single query* running in the per-minute cycle.

As shown in Fig. 8, the LR stream is partitioned "horizontally" over Map nodes; all partitions are *cut* on the same one-minute boundary; the chunk-wise local results are shuffled to the Reduce nodes for global aggregation. The data partition of Map results is based on the standard parallel query processing of "group-by". The system runs cycle by cycle with Map-Reduce applied to data streams in each cycle, hence supporting scaled-out query processing over unbounded data streams.

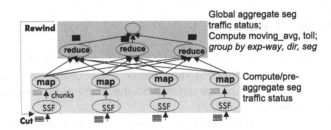

Fig. 8. Parallel DB based streaming map-reduce

This design is being integrated into a commercial parallel database engine where SSF is handled by the storage engine layer at each node, while the Map function and Reduce function are handled by query executers.

6.2 Network-Distributed Map-Reduce Scheme

In network distributed MR scheme, query-engine based stream engines are logically organized in the Map-Reduce style as illustrated in Fig. 9. The separation of "Map" engines and "Reduce" engines are logical, since an engine may act as a "Map" engine, a "Reduce" engine, or both.

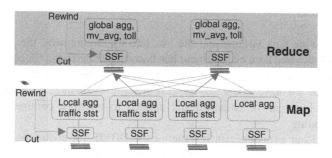

Fig. 9. Network distributed streaming map-reduce

Different from parallel database oriented MR, with the network distributed MR, a specific application is expressed in terms of two cycle based continuous queries, say CQ_{map} and CQ_{reduce}. The same CQ_{map} run at all the Map engines, and the same CQ_{reduce} at all the Reduce engines. The streams are partitioned and fed in multiple CQ_{map}; the resulting streams from CQ_{map} are shuffled to and fused by multiple CQ_{reduce} based on certain grouping criteria specified in the network replicated hash-tables. Those CQ_{map} and CQ_{reduce} synchronized by the same ***cut*** criteria, which determines the boundaries of input streams as well as the resulting streams.

With the above simplified LR example, the stream data are hash partitioned by vehicle ID; the stream data corresponding to express-ways, directions and segments are crossing Map nodes.

- The Map query, CQ_{map}, covers partitioned stream processing, up to the local aggregation of car-volume, speed-sum, group-by time and location.
- The results of CQ_{map} are treated as the input streams of the Reduce query, CQ_{reduce}, partitioned by express-way, direction and segment, based on the network replicated hash tables. Each CQ_{reduce} is also equipped with a SSF for receiving the Map results.
- CQ_{reduce} aggregate segment traffic statistics globally, calculate the segment moving average speed, and then the segment toll.

Both Map and Reduce queries run in the per-minute cycle.

Note the difference CR/MR schemes for parallel DB based and network-distributed MR infrastructure. Since the parallel query engine naturally supports reduce with aggregate-groupby, the MR is expressed by a single query, in each CR cycle the whole MR computation is iterated. With the network distributed MR infrastructure, the Map engines and the Reduce engines run separate cycle-based continuous queries; they process the stream data chunk by chunk based on the common window boundary, or cut criterion, thus cooperate without centralized scheduling. The parallel DB based MR infrastructure generally over-performs the network-distributed one due to efficient data transfer from the Map nodes to the Reduce nodes, but the latter is more flexible and has obvious cost benefits.

7 Conclusions

Due to the growing data volume and the low-latency requirement, the *platform separation* of analytics and data management has become the performance bottleneck, and their integration offers great potential in real-time, data-intensive analytics.

In this work we have addressed several specific challenges. Our thesis is that database technology can be extended and applied to real-time continuous analytics service provisioning.

We reported our experience in leveraging the DBMS for continuous stream analytics. We tackled the key technical issues for integrating stream analytics capability into the existing query engine, and built an integrated, efficient and robust system with stream processing capability while retaining the full DBMS functionality, giving the query engine a new role. We proposed the *cut-and-rewind* query execution model for chunk-wise continuous stream processing with the full SQL power, while enabling history-sensitive stream operations. We provided advanced stream processing capability by extending the existing query engine directly without introducing separate executor or additional "middleware". With this approach we have bridged SQL and stream processing in a single engine.

Our platform significantly differs from the current generation of stream processing systems which are in general built separately from the database systems. As those systems do not have the full SQL expressive power and DBMS functionalities, incur significant overhead in data access and movement, and lack the appropriate transaction support for continuously persisting and sharing results, they fail to meet the requirements for providing high-throughput, low-latency service provisioning.

Further, the cycle-based query model allows multiple query engines to synchronize and cooperate based on the common window boundaries. Such data-driven cooperation is very different from the workflow like centralized scheduling used in other stream processing systems. This feature allows us to apply MR cycle by cycle continuously and incrementally for parallel and distributed continuous analytics, in the way not seen previously. Accordingly, we investigated two kinds of parallel computing infrastructures, one based on parallel database engine; and another based on network distributed Map-Reduce but with extended streaming capability.

The proposed approach has been implemented on the PostgreSQL engine. Our future work includes further refinement of our unified query and transaction model, further characterization and classification of UDFs (to enable optimization) and building out stream analytics operators, additional extensions required for the optimizer and query pipeline, and providing a front-end for demonstrating the live stream analytics. As pointed out in [12], big data visualization issues are tightly coupe with analytics. We are also investigating the use of a massively parallel processor

(MPP)-based, data-intensive streaming analytics platform, and looking into the issues of privacy preservation which plays a critical roles in analytics in both centralized and distributed environments [13, 14].

References

1. Abadi, D., Carney, D., Cetintemel, U., Cherniack, M., Convey, C., Lee, S., Stonebraker, M., Tatbul, N., Zdonik, S.: Aurora a new model and architecture for data stream management. VLDB J. **12**(2), 120–139 (2003)
2. Abadi, D.J., et al.: The design of the borealis stream processing engine. In: CIDR (2005)
3. Arasu, A., B, S., Widom, J.: The CQL continuous query language: semantic foundations and query execution. VLDB J. **15**(2), 121–142 (2006)
4. Bryant, R.E.: Data-intensive supercomputing: the case for DISC. In: CMU-CS-07-128 (2007)
5. Chandrasekaran, S., et al.: TelegraphCQ: continuous dataflow processing for an uncertain world. In: CIDR (2003)
6. Chaiken, R., Jenkins, B., Larson, P.-Å., Ramsey, B., Shakib, D., Weaver, S., Zhou, J.: SCOPE: easy and efficient parallel processing of massive data sets. VLDB **1**(2), 1265–1276 (2008)
7. Chen, J., et al.: NiagaraCQ: a scalable continuous query system for internet databases. In: SIGMOD (2000)
8. Chen, Q., Hsu, M.: Cooperating SQL dataflow processes for In-DB analytics. In: Meersman, R., Dillon, T., Herrero, P. (eds.) OTM 2009, Part I. LNCS, vol. 5870, pp. 389–397. Springer, Heidelberg (2009)
9. Chen, Q., Hsu, M., Liu, R.: Extend UDF technology for integrated analytics. In: Pedersen, T. B., Mohania, M.K., Tjoa, A.M. (eds.) DaWaK 2009. LNCS, vol. 5691, pp. 256–270. Springer, Heidelberg (2009)
10. Cooper, B.F., et al.: PNUTS: Yahoo!'s hosted data serving platform. VLDB. **1**(2), 1277–1288 (2008)
11. Cranor, C.D., et al.: Gigascope: a stream database for network applications. In: SIGMOD (2003)
12. Cuzzocrea, A., Mansmann, S.: OLAP visualization: models, issues, and techniques. In: Wang, J. (ed.) Encyclopedia of Data Warehousing and Mining, 2nd edn, pp. 1439–1446. IGI Global, Hershey (2009)
13. Cuzzocrea, A., Saccà, D.: Balancing accuracy and privacy of OLAP aggregations on data cubes. In: Proceedings of the 13th ACM International Workshop on Data Warehousing and OLAP (DOLAP 2010) in conjunction with 19th ACM International Conference on Information and Knowledge Management (CIKM 2010), Toronto, pp. 93–98, 26–30 October 2010
14. Cuzzocrea, A., Bertino, E.: A secure multiparty computation framework for privacy preserving OLAP over distributed XML data. In: Proceedings of the 25th ACM International Symposium on Applied Computing (SAC 2010), Sierre, pp. 1666–1673, 22–26 March 2010
15. Gedik, B., Andrade, H., Wu, K.-L., Yu, P.S., Doo, M.C.: SPADE: the system s declarative stream processing engine. In: ACM SIGMOD (2008)
16. Franklin, M.J., et al.: Continuous analytics: rethinking query processing in a network-effect world. In: CIDR (2009)

17. Isard, M., Budiu, M., Yu, Y., Birrell, A., Fetterly, D.: Dryad: distributed data-parallel programs from sequential building blocks. In: EuroSys 2007, March 2007
18. Jain, N., et al.: Design, implementation, and evaluation of the linear road benchmark on the stream processing core. In: SIGMOD (2006)
19. Liarou, E., et.al.: Exploiting the power of relational databases for efficient stream processing. In: EDBT (2009)
20. Zeller, H.: NonStop SQL/MX publish subscribe: continuous data streams in transaction processing. In: SIGMOD Conference (2003)

Mining Popular Patterns: A Novel Mining Problem and Its Application to Static Transactional Databases and Dynamic Data Streams

Alfredo Cuzzocrea[1], Fan Jiang[2], Carson K. Leung[2(✉)], Dacheng Liu[2,3],
Aaron Peddle[2], and Syed K. Tanbeer[2]

[1] ICAR-CNR and University of Calabria, Rende, CS, Italy
cuzzocrea@si.dimes.unical.it
[2] University of Manitoba, Winnipeg, MB, Canada
kleung@cs.umanitoba.ca
[3] Wuhan University, Wuhan, Hubei, China

Abstract. Since the introduction of the frequent pattern mining problem, researchers have extended frequent patterns to different useful patterns such as cyclic, emerging, periodic and regular patterns. In this paper, we (i) introduce *popular patterns*, which capture the popularity of individuals, items, or events among their peers or groups. Moreover, we also propose (ii) the *Pop-tree* structure to capture the essential information from transactional databases and (iii) the *Pop-growth* algorithm for mining popular patterns from the Pop-tree. Moreover, we illustrate how our algorithm (iv) mines popular friends from social networks. As we are not confined to mining popular patterns from static transactional databases, we extend our work to mining popular patterns from dynamic data streams. Specifically, we propose (v) the *Pop-stream* structure to capture the popular patterns in batches of data streams and (vi) the *Pop-streaming* algorithm for mining popular patterns from the Pop-stream structure. Experimental results showed that (i) our proposed tree structure is compact and space efficient and (ii) our proposed algorithm is time efficient in mining popular patterns from static transactional databases and dynamic data streams.

Keywords: Data mining · Knowledge discovery · Interesting patterns · Popular patterns · Useful patterns · Tree-based mining · Data streams

1 Introduction

Since the introduction of the research problem of frequent pattern mining, numerous works have been proposed. These works can mostly be classified into two broad "categories". Works in the first "category" mainly focused on algorithmic efficiency, while works in the second "category" mainly focused on extending the notion of frequent patterns to other interesting or useful patterns. However, the mining of these patterns are mostly based on the support/frequency

© Springer-Verlag Berlin Heidelberg 2015
A. Hameurlain et al. (Eds.): TLDKS XXI, LNCS 9260, pp. 115–139, 2015.
DOI: 10.1007/978-3-662-47804-2_6

measure. While support/frequency is a useful metric, support-based frequent pattern mining may *not* be sufficient to discover many interesting knowledge (e.g., popularity) among patterns in a transactional database (TDB). However, in many real-life situations, users want to find popular patterns. For example, a social analyst may want to find persons with large "groups" of friends in social networks as these persons can be the most influential one in their groups or the social networks [9,26]. Similarly, a new member may want to know individuals with high connectivity so that he can get to know more members quickly. A recommender may want to know researchers with large numbers of collaborators. As the fourth example, an event promoter may want to find events with large numbers of participants. With the increase in usage of social network media, it has become more important to be able to find popular individuals (or items, objects, events).

While data in many real-life situations are *static* (e.g., mining popular merchandise items from shopper market basket transactions in a transactional database), the automation of measurements and data collection in some other real-life situations is producing *dynamic* streams of data. For instance, the development and increasing use of a large number of sensors (e.g., acoustic, chemical, electromagnetic, mechanical, optical radiation and thermal sensors) for various real-life applications have led to data streams [11,23].

Hence, in this paper, we aim to mine popular patterns from both static transactional databases and dynamic data streams. Specifically, our **key contributions** of this paper include the following:

1. our introduction of the notion of *popular patterns*;
2. our proposal of the *Pop-tree*, which is a tree structure to capture essential information about the popularity of individuals, items, objects, or events;
3. our design and development of the *Pop-growth algorithm*, which mines popular patterns from the Pop-tree capturing static data;
4. our proposal of the *Pop-stream*, which is a structure to capture popular individuals, items, objects, or events mined from batches of dynamic data streams; and
5. our design and development of the *Pop-streaming algorithm*, which finds popular patterns from the Pop-stream structure capturing dynamic data.

As the current paper is an extension and enhancement of our DaWaK 2012 paper [25], additional contributions beyond the basic mining of popular patterns from static transactional databases include the following: (i) extending the mining of popular patterns from static transactional databases to the mining of popular patterns from dynamic data streams (Sect. 6), (ii) demonstration of our algorithm for a useful application of mining popular friends from social networks (Sect. 5), and (iii) running additional experiments (especially for popular pattern mining from streams).

The remainder of this paper is organized as follows. The next section reviews some related works. We introduce popular patterns in Sect. 3. In Sect. 4, we propose (i) the Pop-tree structure that captures important contents of the TDB and

(ii) the Pop-growth algorithm that constructs the Pop-tree, from which popular patterns can be mined recursively. In Sect. 6, we propose (i) the Pop-stream structure that captures popular patterns mined from batches of data streams and (ii) the Pop-streaming algorithm that calls Pop-growth to find popular patterns from a batch, stores the mined patterns in the Pop-stream structure, and returns to users popular patterns that can be mined from the dynamic data streams. We demonstrate a useful real-life application of mining popular friends from social networks in Sect. 5. Experimental results are presented in Sect. 7. Finally, conclusions and future work are provided in Sect. 8.

2   Related Work

Recall from the previous section that numerous works have been proposed since the introduction of the research problem of frequent pattern mining. These works can mostly be classified into two broad "categories". Works in the first "category" mainly focused on algorithmic efficiency [22,23]. For example, to avoid the candidate generation-and-test approach of the Apriori algorithm [1], a tree-based algorithm called FP-growth [17] was proposed to build an FP-tree to capture the contents of a TDB so that frequent patterns can be mined recursively from the FP-tree with a restricted test-only approach.

Works in the second "category" mainly focused on extending the notion of frequent patterns to other interesting or useful patterns such as sequences, episodes, maximal and closed sets. Note that the mining of these patterns are mostly based on the support/frequency measure. While support/frequency is a useful metric, support-based frequent pattern mining may *not* be sufficient to discover many interesting knowledge (e.g., correlation, regularity, periodicity, popularity) among patterns in a TDB. This leads to the introduction of some interestingness measures [31] and their corresponding patterns such as emerging patterns [2], constrained patterns [19,24], correlated patterns [20], periodic patterns [29,34], regular patterns [30], hyperclique patterns [32], and high utility patterns [33].

Nowadays, the automation of measurements and data collection is producing tremendously huge volumes of data. For instance, the development and increasing use of a large number of sensors (e.g., acoustic, chemical, electromagnetic, mechanical, optical radiation and thermal sensors) for various real-life applications (e.g., environment surveillance, security, manufacturing systems) have led to data streams [11,23]. To discover useful knowledge from these streaming data, several mining algorithms [10,12,14] have been proposed. In general, mining frequent patterns from dynamic data streams [16,18,28] is more challenging than mining from traditional static transaction databases due to the following characteristics of data streams:

1. Data streams are continuous and unbounded. As such, we no longer have the luxury to scan the streams multiple times. Once the streams flow through, we lose them. We need some techniques to capture important contents of the

streams. For instance, landmark windows capture contents of all batches after the landmark (i.e., sizes of windows keep increasing with the number of batches).

2. Data in the streams are not necessarily uniformly distributed. As such, a currently infrequent pattern may become frequent in the future and vice versa. We have to be careful not to prune infrequent patterns too early; otherwise, we may not be able to get complete information such as frequencies of some patterns (as it is impossible to recall those pruned patterns).

Over the past few years, several stream mining algorithms—including FP-streaming [15], UF-streaming [22], TUF-streaming [21,23], and XTUF-streaming [21]— have been proposed. However, most of them mine *frequent* patterns (instead of popular patterns. In contrast, our Pop-streaming algorithm mines *popular* patterns.

3   Our Proposed Notion of Popular Patterns

Let Item=$\{x_1, x_2, \ldots, x_m\}$ be a set of m domain items. A transactional database (TDB) is the set of n transactions: $\{t_1, t_2, \ldots, t_n\}$, where each transaction t_j in the TDB is a subset of Item. We use $|t_j|$ to represent the transaction length of t_j. Let $X = \{x_1, x_2, \ldots, x_k\} \subseteq$ Item be a pattern consisting of k items (i.e., a k-itemset), where $|X| = k \leq m$. Then, the projected database of X (denoted as DB_X) is a set of TDB transactions (in the TDB) that contain X. We use $maxTL(X)$ and $sumTL(X)$ to respectively represent the maximum length and the total length of all transactions in DB_X.

Example 1. Consider the TDB shown in Table 1, which consists of n=7 transactions and m=9 domain items $a, b, \ldots, i$. For pattern $X = \{b, c\}$, its projected database $DB_{\{b,c\}} = \{t_2, t_3\}$. Hence, $|DB_{\{b,c\}}| = 2$. In other words, the support (or frequency) of $\{b, c\}$ in the TDB is $sup(\{b, c\}, \text{TDB}) = sup(\{b, c\}, DB_{\{b,c\}}) = 2$. Moreover, $|t_2| = |\{b, c, f, g, h\}| = 5$, $|t_3| = |\{b, c, d, e, f, h\}| = 6$, $maxTL(\{b, c\}) = \max\{|t_2|, |t_3|\} = \max\{5, 6\} = 6$, and $sumTL(\{b, c\}) = |t_2| + |t_3| = 5 + 6 = 11$. □

Table 1. A transaction database

Transaction ID	Transaction
t_1	$\{b, d\}$
t_2	$\{b, c, f, g, h\}$
t_3	$\{b, c, d, e, f, h\}$
t_4	$\{c, e, g, h\}$
t_5	$\{a, d\}$
t_6	$\{a, b, i\}$
t_7	$\{a, d, e\}$

Definition 1. *The* **transaction popularity** $Pop(X, t_j)$ *of a pattern X in transaction t_j measures the membership degree of X in t_j. For simplicity, we compute the membership degree based on the difference between the transaction length $|t_j|$ and the pattern size $|X|$:*

$$Pop(X, t_j) = |t_j| - |X|. \tag{1}$$

Note that, depending on real-life applications, the above equation can be adapted to incorporate some other functional operations on t_j and X.

Definition 2. *The* **long transaction popularity** $Pop(X, t^{maxTL(X)})$ *of a pattern X in transaction $t^{maxTL(X)}$ measures the membership degree of X in $t^{maxTL(X)}$, where $t^{maxTL(X)}$ is the transaction having the maximum length in DB_X:*

$$Pop(X, t^{maxTL(X)}) = \left(\max_{t_j \in DB_X} |t_j| \right) - |X|. \tag{2}$$

Definition 3. *The* **popularity** $Pop(X)$ *of a pattern X in DB_X measures an aggregated membership degree of X in all transactions in DB_X. It is defined as an average of all transaction popularities of X:*

$$Pop(X) = \frac{1}{sup(X, DB_X)} \sum_{t_j \in DB_X} Pop(X, t_j). \tag{3}$$

Example 2. Reconsider the TDB shown in Table 1. The transaction popularity of pattern $\{b, c\}$ in t_2 can be computed as $Pop(\{b, c\}, t_2) = |t_2| - |\{b, c\}| = 5 - 2 = 3$. Similarly, $Pop(\{b, c\}, t_3) = |t_3| - |\{b, c\}| = 6 - 2 = 4$. Recall from Example 1 that $DB_{\{b,c\}} = \{t_2, t_3\}$ (i.e., $\{b, c\}$ appears only in t_2 and t_3). As t_3 is the longest transaction in $DB_{\{b,c\}}$ (because $maxTL(\{b, c\}) = 6$), the long transaction popularity of pattern $\{b, c\}$ in $t^{maxTL(\{b,c\})}$ can be computed as $Pop(\{b, c\}, t^{maxTL(\{b,c\})}) = \max\{|t_2|, |t_3|\} - |\{b, c\}| = 6 - 2 = 4$. Hence, the popularity of pattern $\{b, c\}$ is $\frac{1}{sup(\{b,c\}, DB_{\{b,c\}})}(Pop(\{b, c\}, t_2) + Pop(\{b, c\}, t_3)) = \frac{1}{2}(3+4) = 3.5$ □

Definition 4. *Given a user-specified minimum popularity threshold minpop, a pattern X is considered* **popular** *if its popularity is at least minpop (i.e., $Pop(X) \geq minpop$).*

Example 3. If the user-specified *minpop* is 3.3, then pattern $\{b, c\}$ is popular in the TDB shown in Table 1 because $Pop(\{b, c\}) = 3.5 \geq 3.3 = minpop$. However, pattern $\{b\}$ is *not* popular because $Pop(\{b\}) = \frac{1}{sup(\{b\}, DB_{\{b\}})}(Pop(\{b\}, t_1) + Pop(\{b\}, t_2) + Pop(\{b\}, t_3) + Pop(\{b\}, t_6)) = \frac{1}{4}(1+4+5+2) = 3 < 3.3 = minpop.$ □

4   Pop-Growth: Mining Popular Patterns from Static Transactional Databases

When mining frequent patterns, the frequency measure satisfies the downward closure property (i.e., if a pattern is infrequent, its superset is guaranteed to be infrequent). This helps reduce the search/solution space by pruning infrequent patterns, and thus speeds up the mining process. However, when mining popular patterns, observant readers may notice from Example 3 that popularity does *not* satisfy the downward closure property. For example, a pattern (e.g., $\{b\}$) is unpopular, but its superset (e.g., $\{b, c\}$) may be popular. Hence, the mining of popular patterns can be challenging.

To handle the challenge, let us revisit Eq. (3) and redefine the popularity $Pop(X)$ of a pattern X (cf. Definition 3).

Definition 5. *The* **popularity** $Pop(X)$ *of a pattern X in DB_X measures an aggregated membership degree of X in all transactions in DB_X. It is defined in terms of $sumTL(X) = \sum_{t_j \in DB_X} |t_j|$ as follows:*

$$
\begin{aligned}
Pop(X) &= \frac{1}{sup(X, DB_X)} \sum_{t_j \in DB_X} Pop(X, t_j) \\
&= \frac{1}{sup(X, DB_X)} \sum_{t_j \in DB_X} (|t_j| - |X|) \\
&= \frac{sumTL(X)}{sup(X, DB_X)} - |X|.
\end{aligned}
\tag{4}
$$

Example 4. Reconsider the TDB shown in Table 1. Recall from Example 1 that $sumTL(\{b,c\})=11$. Then, the popularity of pattern $\{b, c\}$ is $\frac{sumTL(\{b,c\})}{|\{t_2,t_3\}|} - |\{b,c\}| = \frac{11}{2} - 2 = 3.5$. Similarly, the popularity of pattern $\{b\}$ is $\frac{sumTL(\{b\})}{|\{t_1,t_2,t_3,t_6\}|} - |\{b\}| = \frac{|t_1|+|t_2|+|t_3|+|t_6|}{|\{t_1,t_2,t_3,t_6\}|} - |\{b\}| = \frac{16}{4} - 1 = 3$.   □

Observant readers may notice from Example 4 that $sumTL(\{b,c\})=11 \leq 16=sumTL(\{b\})$. The definition of $sumTL(X)$ further confirms that the total transaction length $sumTL(X)$ of X satisfies the downward closure property. In other words, for $X \subseteq X'$,

$$
sumTL(X) \geq sumTL(X').
\tag{5}
$$

4.1 Construction of a Pop-Tree

To mine popular patterns, we propose the Pop-growth algorithm, which consists of two key procedures: (i) construction of a Pop-tree and (ii) mining of popular patterns from the Pop-tree.

We first build a tree structure—called **Popular pattern tree (Pop-tree)**—to capture the necessary information from the TDB with only two scans of the TDB. Recall from Sect. 3 that $Pop(X)$ does not satisfy the downward closure property. So, unpopular items need to be kept in the Pop-tree as some of their supersets may be popular. Fortunately, recall from Sect. 3 that $sumTL(X)$ satisfies the downward closure property. So, not all unpopular items need to be kept. Some of them can be pruned. See the following two lemmas.

Lemma 1. *The popularity of a pattern X is always less than or equal to its long transaction popularity, i.e., $Pop(X) \leq Pop(X, t^{maxTL(X)})$.*

Proof. Recall from Equation (3) that $Pop(X) = \frac{1}{sup(X,DB_X)} \sum_{t_j \in DB_X} Pop(X, t_j)$, where $Pop(X, t_j) = |t_j| - |X|$. According to Equation (1), $Pop(X, t_j)$ measures the membership degree of X in t_j. As shown in Equation (2), The long transaction popularity $Pop(X, t^{maxTL(X)})$ measures the membership degree of X in the longest transaction containing X. Hence, $Pop(X)$ is always less than or equal to $Pop(X, t^{maxTL(X)})$:

$$Pop(X) = \frac{1}{sup(X, DB_X)} \sum_{t_j \in DB_X} Pop(X, t_j)$$

$$= \frac{1}{sup(X, DB_X)} \sum_{t_j \in DB_X} (|t_j| - |X|)$$

$$= \frac{\sum_{t_j \in DB_X} |t_j|}{|DB_X|} - |X|$$

$$= \left(\underset{t_j \in DB_X}{avg} |t_j| \right) - |X|$$

$$\leq \left(\underset{t_j \in DB_X}{max} |t_j| \right) - |X| = Pop(X, t^{maxTL(X)}). \qquad (6)$$

Hence, this proved that $Pop(X) \leq Pop(X, t^{maxTL(X)})$. $\qquad\qquad\qquad\square$

Lemma 2. *For $X \subseteq X'$, $Pop(X')$ cannot exceed $maxTL(X) - |X'|$.*

Proof. Recall from Equation (4) that $Pop(X) = \frac{sumTL(X)}{sup(X,\text{TDB})} - |X|$. Knowing that $sumTL(X)$ satisfies the download closure property, we get $sumTL(X') \leq sumTL(X)$ for $X \subseteq X'$ as shown in Equation (5). Then, we get the following:

$$Pop(X') = \frac{sumTL(X')}{sup(X', \text{TDB})} - |X'|$$

$$\leq maxTL(X') - |X'|$$

$$\leq maxTL(X) - |X'| \qquad (7)$$

Hence, this proved that $Pop(X') \leq maxTL(X) - |X'|$. $\qquad\qquad\qquad\square$

Based on the above two lemmas, the following equation provides us with an upper bound of the popularity $Pop(X')$ of a pattern X' (in terms of $maxTL(X)$), where $X \subseteq X'$:

$$Pop^{UB}(X') \leq maxTL(X) - |X'|. \qquad (8)$$

Based on Equation (8), we can first calculate the popularity upper bound of a pattern X' from $maxTL(X)$ where (i) $X \subseteq X'$, and (ii) $|X'| \geq |X| + 1 = k + 1$). We can then prune unpopular patterns. We call this *super-pattern popularity check*.

Similar to FP-tree [17], each node of a Pop-tree contains the parent and child pointers as well as horizontal node traversal pointers. To facilitate popular pattern mining, we keep (i) an item x, (ii) support of $Y \cup \{x\}$, (iii) $sumTL(Y \cup \{x\})$, and (iv) $maxTL(Y \cup \{x\})$, where Y represents the set of items above x (i.e., ancestor nodes of x).

To construct a Pop-tree, we scan the TDB to find the $support(x)$, maximum transaction length $maxTL(x)$ and the popularity $Pop(x)$ for each singleton x in the TDB. Then, we perform the super-pattern popularity check and safely delete a pattern x if $Pop^{UB}(x') < minpop$ (where x' is an extension of x). We then scan the TDB the second time to insert each transaction into the Pop-tree in a similar fashion as the insertion process of FP-tree.

Example 5. Let us show how to construct a Pop-tree for the TDB shown in Table 1 with $minpop$=2.4. With the first database scan, we obtain the following information in the form of

$$\langle x: sup(x, \text{TDB}), maxTL(x), Pop(x) \rangle$$

for each of the m=9 domain items: $\langle a{:}3,3,1.66 \rangle$, $\langle b{:}4,6,3.0 \rangle$, $\langle c{:}3,6,4.0 \rangle$, $\langle d{:}4,6,2.25 \rangle$, $\langle e{:}3,6,3.33 \rangle$, $\langle f{:}2,6,4.5 \rangle$, $\langle g{:}2,5,3.5 \rangle$, $\langle h{:}3,6,4.0 \rangle$, $\langle i{:}1,3,2.0 \rangle$. This information is useful as follows. (i) Based on the obtained $Pop(x)$ values, we noticed that all items—except a, d &i—are popular (i.e., with popularity at least 2.4). Although $\{a\}$, $\{d\}$ &$\{i\}$ are unpopular, their super-patterns may be popular. Hence, we cannot delete them without performing the super-pattern popularity check. So, (ii) the obtained $maxTL(x)$ values are used for super-pattern popularity check. The popularity upper bounds of the extensions of $\{a\}$, $\{d\}$ &$\{i\}$ are at most $3-2=1$, $6-2=4$ &$3-2=1$, respectively. As the value for $Pop^{UB}(\{d\})$ is greater than $minpop$, we keep $\{d\}$ but safely delete $\{a\}$ &$\{i\}$. Finally, (iii) we sort and insert items b, c, d, e, f, g &h into a header table (H-table) in the descending order of the obtained $sup(x, \text{TDB})$ values: $\langle b, d, c, e, h, f, g \rangle$.

We then scan the TDB the second time. We compute the length of each transaction, remove all items that are not in the H-table, and sort the remaining

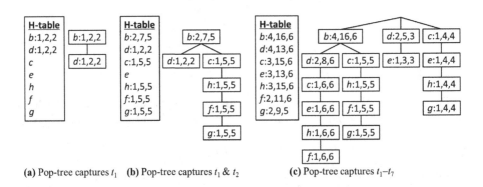

(a) Pop-tree captures t_1 (b) Pop-tree captures t_1 & t_2 (c) Pop-tree captures t_1–t_7

Fig. 1. The Pop-tree construction

items in each transaction according to the H-table order. Figure 1(a) shows the contents of the H-table in the form of

$$\langle x\colon sup(x, \text{TDB}),\ sumTL(x),\ maxTL(x)\rangle,$$

and the Pop-tree structure after inserting t_1 of TDB. Because t_1 and t_2 share a common prefix (i.e., $\{b\}$), we increase the occurrence count of the common node $\{b : 1, 2, 2\}$ by one, its total transaction length ($sumTL$) by the transaction length of t_2 (i.e., $|t_2|=5$), and update its $maxTL$. For the remaining (uncommon) nodes of t_2, we set $support=1$, $sumTL=|t_2|$ and $maxTL=|t_2|$. The contents of the Pop-tree after insertion of t_2 are shown in Fig. 1(b). The final Pop-tree after capturing all the transactions in the TDB is shown in Fig. 1(c).   □

Let $I(t_j)$ be the set of items in transaction t_j that pass through the first database scan. Based on the above Pop-tree construction procedure, we observed several important properties of Pop-trees listed as follows.

Property 1. A Pop-tree registers the projection of $I(t_j)$ for t_j in the TDB only once.

Property 2. The total transaction length $sumTL$ in a node x in a Pop-tree captures the sum of lengths of all transactions that pass through, or end at, the node for all the nodes in the path from x up to the root.

Property 3. The total transaction length $sumTL$ of any node in a Pop-tree is greater than or equal to the sum of transaction lengths of its children.

Properties 2 and 3 are the result of sharing common prefixes by different transactions, which allow our Pop-tree to be compact. Based on following lemma, one can observe that a Pop-tree is a highly compact tree structure.

Lemma 3. *The size of a Pop-tree on a TDB for minpop is bounded above by* $\sum_{t_j \in \text{TDB}} |I(t_j)|$.

Proof. Recall that $I(t_j)$ denotes the set of items in transaction t_j that pass through the first database scan. In other words, $I(t_j)$ represents the set of individually popular items in t_j. During the Pop-tree construction, these items in t_j are inserted as a tree path into a Pop-tree for popular pattern mining. Thus, as we insert every transaction t_j in the TDB, the size of the resulting Pop-tree—in terms of the total number of tree nodes—would be equal to the total number of individually popular items in all the transactions, i.e., $\sum_{t_j \in \text{TDB}} |I(t_j)|$. This would be the worst case scenario, in which there is no tree path sharing (i.e., no common nodes can be merged). Fortunately, in most cases, some paths are in common and can thus be merged. In those cases, the size of a Pop-tree—in terms of the total number of tree nodes—would be lower than the total number of individually popular items in all transactions. Hence, this proved that the size of a Pop-tree on a TDB for *minpop* is bounded above by $\sum_{t_j \in \text{TDB}} |I(t_j)|$.   □

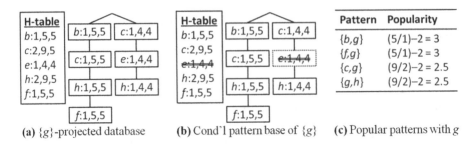

Fig. 2. The Pop-tree mining

Lemma 4. *Given a TDB and minpop, the complete set of all popular patterns can be obtained from a Pop-tree for the minpop on the TDB.*

Proof. Given a TDB and *minpop*, the Pop-tree keeps every item x with $Pop(x) \geq$ *minpop* (i.e., keeps all individually popular items). Every possible pattern X with $Pop^{UB}(X) \geq$ *minpop* can then be generated from the Pop-tree. As $Pop^{UB}(X) \geq Pop(X)$ for any pattern X, this implies that all patterns X with $Pop(X) \geq$ *minpop* (i.e., true positives) can be generated. As a by-product, some patterns Y with $Pop^{UB}(Y) \geq$ *minpop* $> Pop(Y)$ (i.e., false positives) could be generated from the Pop-tree. As a preview, the corresponding Pop-growth algorithm would prune these false positives as its last step and would return only those true positives to the user. Hence, given a TDB and *minpop*, the complete set of all popular patterns can be obtained from a Pop-tree for the minpop on the TDB. $\square$

We can justify the completeness of a Pop-tree for mining popular patterns by Lemma 4. Based on this lemma, popular patterns can be found by mining our Pop-tree.

4.2   Finding Popular Patterns from the Pop-Tree

Recall that, to mine popular patterns, the Pop-growth algorithm applies two key procedures: (i) construction of a Pop-tree and (ii) mining of popular patterns from the Pop-tree. The Pop-growth finds popular patterns from the Pop-tree, in which each tree node captures its occurrence count, total transaction length, and maximum transaction length. The algorithm finds popular patterns by constructing the projected database for potential popular patterns and recursively mining their extensions.

While constructing the conditional pattern base from a projected database, we perform a super-pattern popularity check for extensions of any unpopular item, and delete the item only when it fails the check. We call such pruning technique the *lazy pruning*.

The lazy pruning technique ensures that no popular patterns (having unpopular subsets) will be missed by Pop-growth. The following example illustrates how Pop-growth mines popular patterns from the Pop-tree.

Example 6. Let us continue Example 5. In other words, let us mine popular patterns from the Pop-tree shown in Fig. 1(c) constructed for the TDB shown in Table 1 with *minpop*=2.4.

Recall that the Pop-growth recursively mines the projected databases of all items in H-table. Before constructing the projected database for an item x in H-table, we output the item as a popular pattern if its popularity is at least *minpop*. The conditional pattern base for the $\{g\}$-projected database (i.e., $DB_{\{g\}}$), as shown in Fig. 2(a), is constructed by accumulating the contents in the tree paths: $\langle b{:}1,5,5 \quad c{:}1,5,5 \quad h{:}1,5,5 \quad f{:}1,5,5 \rangle$ and $\langle c{:}1,4,4 \quad e{:}1,4,4 \quad h{:}1,4,4 \rangle$. The header table for $DB_{\{g\}}$, as shown in Fig. 2(a), contains all items that co-occur with g in the Pop-tree. It also contains the corresponding *support, sumTL* and *maxTL* of each item in $DB_{\{g\}}$. We then compute the exact popularity of each item in $DB_{\{g\}}$ by using Eq. (4).

The conditional tree for any conditional pattern base of a pattern X may contain two types of items: (i) items that are popular in DB_X and (ii) items that are unpopular in DB_X but have potentially popular super-patterns. Other items are deleted from the projected database. To find unpopular items that having potentially popular super-patterns, we apply the lazy pruning technique and Eq. (8).

Based on Eq. (4), the popularity of items in the H-table of $DB_{\{g\}}$ can be computed: $Pop(\{b,g\}) = \frac{5}{1} - 2 = 3$, $Pop(\{f,g\}) = \frac{5}{1} - 2 = 3$, $Pop(\{c,g\}) = \frac{9}{2} - 2 = 2.5$, $Pop(\{g,h\}) = \frac{9}{2} - 2 = 2.5$ and $Pop(\{e,g\}) = \frac{4}{1} - 2 = 2$. All items except e are popular together with g. By applying the lazy pruning technique, the popularity upper bound $Pop^{UB}(\{e,g\})$ for e with g can be calculated as at most $4-2 = 2$, which is less than *minpop*. Hence, we can safely delete e from the projected database of g. The conditional tree for the projected database of g is presented in Fig. 2(b).

The mining for each extension (i.e., for f, h, c &b) of g is performed recursively. The set of patterns generated from the projected database of g is shown in Fig. 2(c). The mining process terminates when we reach the top of H-table of the Pop-tree.   □

The Pop-growth mining technique is efficient because it applies a pattern-growth based mining technique on a Pop-tree. Moreover, the lazy pruning technique further reduces the mining cost for unpopular items whose super-patterns cannot be popular.

5   Discussion: An Application on Mining Popular Friends from Social Networks

In the previous section, we described how our proposed Pop-growth mines frequent patterns from transactional databases. This algorithm builds a Pop-tree structure to capture important contents of the transactional databases and recursively mines popular patterns from the Pop-tree. In this section, we extend the proposed technique and apply it to mine a special type of frequent patterns—popular friends—from social networks.

Recent advances in technology and successes in online digital media sites have led the surge of interest in social computing and its applications. Social computing enables users to intersect social behaviour with computing systems and to create social conventions as well as social contexts through the use of software and technology. This explains why, over the past few years, various research works on the analytics, mining and visualization of complex social networks have been proposed. In general, social networks are structures made of social entities (e.g., individuals, corporations, collective social units, or organizations) that are linked by some specific types of interdependencies (e.g., kinship, friendship, common interest, beliefs, or financial exchange). These dependencies among linked entities in the social networks present an opportunity to further infer different properties of individuals. Because a social entity is connected to another social entity as his next-of-kin, friend, collaborator, co-author, classmate, co-worker, team member and/or business partner, identifying social entities or groups of entities that have connections with a large number of other social entities may provide useful knowledge to the user. For example, among the friends of p, some of them may be very popular in the sense that they have many connections. Discovering these popular friends provides useful knowledge to p because they may have high social connectivity and/or could have strong influence on members of their social groups. Similarly, a newcomer (to a city, company, or profession) may want to be introduced to individuals having high social connectivity so that he can get to know more people faster. Similar comments apply to users in other social networking sites.

Note that the task of finding popular friends can be more complicated when we do not have access to these lists of connections. For example, due to various reasons (e.g., privacy setting), connection lists of some social entities in the social network may not be accessible to unauthorized users. Fortunately, members of interest groups (especially, open public groups) are usually visible. In these situations, given a social network containing all members of these interest groups, we can adapt our proposed Pop-growth to find popular users or a popular group of friends. Specifically, to adapt our proposed Pop-growth algorithm for mining popular friends from social networks, we treat (i) each interest group list like a transaction and (ii) each social user/member in an interest group list like an item in a transaction. With this adaption and setting of a user-specified *minpop*, we find popular friends from a sample social network as illustrated in Example 7.

Example 7. Consider a collection of $n=7$ interest groups involving $m=9$ users (namely, Alice, Bob, Cathy, Don, Ed, Fank, Gary, Helen, and Irene) as shown in Table 2, which may represent a subset of a large social network. To adapt our proposed Pop-growth algorithm for mining popular friends from social networks, we treat (i) each interest group list like a transaction and (ii) each social user/member in an interest group list like an item. With this adaption and setting of a user-specified *minpop*=2.4, we first scan the collection once to find *individually popular users*: Bob, Cathy, Ed, Frank, Gary, or Helen, with their popularity values of 3, 4, 3.33, 4.5, 3.5, or 4, respectively.

Table 2. A sample social network

Interest group list ID	Members in the interest group list
L_1 on ballet	{Bob, Don}
L_2 on baseball	{Bob, Cathy, Frank, Gary, Helen}
L_3 on curling	{Bob, Cathy, Don, Ed, Frank, Helen}
L_4 on football	{Cathy, Ed, Gary, Helen}
L_5 on hockey	{Alice, Don}
L_6 on lacrosse	{Alice, Bob, Irene}
L_7 on soccer	{Alice, Don, Ed}

Note that, among the $m=9$ users, only six of them are popular. As for the three unpopular users (Alice, Don, and Irene), their super-pattern may still be popular. Hence, we cannot delete them without performing the super-pattern popularity checks. The checks reveal that, when $X=${Alice}, the popularity upper bound $Pop^{UB}(X')$ of its extension X' (where $X' \supseteq$ {Alice}; e.g., X' = {Alice, Bob}) is at most $3-2=1 < minpop$. Similarly, when $X=${Irene}, the popularity upper bound $Pop^{UB}(X')$ of its extension X' is also at most $3-2=1 < minpop$. In contrast, when $X=${Don}, the popularity upper bound $Pop^{UB}(X')$ of its extension X' is at most $6-2=4 \geq minpop$. Hence, we keep Don but safely delete Alice and Irene.

To find popular groups of users, we build a Pop-tree by sorting and inserting the six individual popular users (Bob, Cathy, Ed, Frank, Gary, and Helen) together with this undeleted user (Don) into a header table (H-table) in descending order of their support values: ⟨Bob, Don, Cathy, Ed, Helen, Frank, Gary⟩. We then scan the collection of $n=7$ interest group lists the second time. During that, we compute the length of each interest group list (e.g., $|L_1| = 2, |L_3| = 6$), remove all uses who are not in the H-table (e.g., remove Alice from L_5 to make the resulting list become {Don}, remove both Alice and Irene from L_6 to make it become {Bob}, remove Alice from L_7 to make it become {Don, Ed}), and sort the remaining users in each interest group list according to the H-table order. When inserting users into a Pop-tree, if two interest group lists share a common prefix (e.g., Bob appears in both L_1 and L_2), then the prefix is merged. Figure 3 shows the Pop-tree after capturing all the interest group lists in the social collection.

Once the Pop-tree is built, we call Pop-growth to recursively mine the projected databases of all users in H-table. Before constructing the projected database for a user (e.g., {Gary}) in H-table, we output it as a popular user if its popularity is at least $minpop$. The conditional pattern base for the {Gary}-projected database (i.e., $DB_{\{Gary\}}$) is constructed by accumulating the contents in the tree path ⟨Bob:1,5,5 Cathy:1,5,5 Helen:1,5,5 Frank:1,5,5⟩ and ⟨Cathy:1,4,4 Ed:1,4,4 Helen:1,4,4⟩. The header table for $DB_{\{Gary\}}$ contains all users that share a common interest with Gary in the Pop-tree. It also contains the

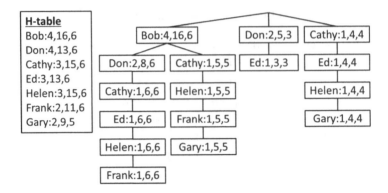

Fig. 3. The Pop-tree capturing interest group lists of a sample social network

corresponding *support*, *sumTL* and *maxTL* of each user in $DB_{\{Gary\}}$. We then compute the exact popularity of each item in $DB_{\{Gary\}}$ by using Eq. (4).

As the conditional tree for any conditional pattern base of a group X of users contains (i) those who are popular in DB_X and (ii) those who are unpopular in DB_X but have potentially popular super-groups, we apply the lazy pruning technique and Eq. (8) to prune out those unpopular users who have potentially popular super-groups. For example, we found popular groups {Bob, Gary} and {Frank, Gary} (both with popularity value of 3), as well as {Cathy, Gary} and {Gary, Helen} (both with popularity value of 2.5). Similar steps can be applied to other paths in a Pop-tree to find all other popular friends or groups of friends from social networks.   □

6   Pop-Streaminng: Mining Popular Patterns from Dynamic Data Streams

In Sect. 4, we mined popular patterns from static transactional databases. The corresponding Pop-growth algorithm works well when handling static data. However, there are situations in which we need to deal with dynamic streaming data. In this section, we propose another algorithm for handling dynamic data. The corresponding algorithm—called **Pop-streaming**—mines popular patterns from dynamic data streams in a *landmark model* environment.

When using the landmark model for processing data streams, transactions in each batch (regardless of whether they are historical or recent data) are treated equally. As such, all batches (regardless of whether they are old or recent) are assigned the same weights. To mine popular patterns from dynamic data streams, our proposed Pop-streaming algorithm first calls Pop-growth (Sect. 4) to find popular patterns from the current batch of transactions in the streams (using a threshold called *preMinpop*, which is defined to be ≤ *minpop*). Note that, although users are interested in truly popular patterns (i.e., patterns with popularity ≥ *minpop* > *preMinpop*), *preMinpop* is used in attempt to avoid pruning a pattern too early. This is important because data in the continuous streams are not necessarily uniformly distributed.

Once the popular patterns for a batch of streaming data are found, the next step is to construct a **Pop-stream structure** to capture the mined popular patterns. Each node in this tree-based Pop-stream structure corresponds to a popular pattern. Nodes that correspond to the popular patterns sharing common prefix are merged. In addition to the popular pattern X (or more precisely, the suffix item in X), each node stores additional information. So, on the surface, this Pop-stream structure may seem to be similar to that of the UF-stream structure [21] used in frequent pattern mining. As such, it was tempting to keep X and its popularity value $Pop(X)$ in each node. However, a closer look reveals that, while frequency (or support) of patterns is additive, popularity of patterns is *not*. See Example 8.

Table 3. A data stream

Batch ID	Transaction ID	Transaction
B_1	t_1	$\{b, d\}$
	t_2	$\{b, c, f, g, h\}$
B_2	t_3	$\{b, c, d, e, f, h\}$
	t_4	$\{c, e, g, h\}$

Example 8. Consider two batches of streaming data as shown in Table 3. The support of $\{c, h\}$ in Batch B_1 is 1, and that in Batch B_2 is 2. So, $sup(\{c, h\}, B_1 \cup B_2) = sup(\{c, h\}, B_1) + sup(\{c, h\}, B_2) = 1 + 2 = 3$. However, the popularity of $\{c, h\}$ in Batch B_1 is $\frac{1}{|\{t_2\}|}Pop(\{c, h\}, t_2) = 3$, and that in Batch B_2 is $\frac{1}{|\{t_3, t_4\}|}(Pop(\{c, h\}, t_3) + Pop(\{c, h\}, t_4)) = \frac{1}{2}(4 + 2) = 3$. So, the sum of these two popularity values becomes $3+3 = 6$, which is *not* equal to the popularity of $\{c, h\}$ in these two batches. Mathematically, $Pop(\{c, h\}, B_1 \cup B_2) = \frac{1}{|\{t_2, t_3, t_4\}|}(Pop(\{c, h\}, t_2) + Pop(\{c, h\}, t_3) + Pop(\{c, h\}, t_4)) = \frac{1}{3}(3+4+2) = 3.\square$

Recall from Definition 5 that the popularity $Pop(X)$ of a pattern X can be computed in terms of (i) $sumTL(X)$ and (ii) $sup(X, B_i)$. Moreover, both (i) $sumTL(X)$ and (ii) $sup(X, B_i)$ are additive. For example, $sumTL(\{c, h\})$ in B_1 is 5, $sumTL(\{c, h\})$ in B_2 is $6+4 = 10$, whereas $sumTL(\{c, h\})$ in the first two batches is $5+(6+4) = 15$. Similarly, $sup(\{c, h\}, B_1)$ is 1, $sup(\{c, h\}, B_2)$ is 2, whereas $sup(\{c, h\}, B_1 \cup B_2)$ is $1+2 = 3$. Hence, instead of storing the popularity value of a popular pattern, we store (i) $sumTL(X)$ and (ii) $sup(X, B_i)$ values so that we can compute the popularity of X based on these two values.

As we are dealing with batches of streaming data, the Pop-stream structure needs to be updated. Hence, we need to store multiple pairs of $sumTL(X)$ and $sup(X, B_i)$ values (one pair for each batch) In other words, we need to store w pairs of $sumTL(X)$ and $sup(X, B_i)$ values when handling w batches of streaming data Fortunately, when using the landmark model, all data are of the same weights. Hence, we only need keep *one* pair of $sumTL(X)$ and $sup(X, B_i)$ values for each node representing a popular pattern X:

$$(i) \ X, \ (ii) \ sumTL(X), \ \text{and} \ (iii) \ sup(X, \textstyle\bigcup_i B_i).$$

When a new batch B_j flows in, if X does not exist in the Pop-stream structure, our Pop-streaming algorithm inserts $\langle X, sumTL(X), sup(X, B_j) \rangle$ into the Pop-stream structure. Otherwise (i.e., X exists in the Pop-stream structure), we need to update the stored information as follows:

1. add the new $sumTL(X)$ to the existing $sumTL(X)$, and
2. add the new $sup(X, B_j)$ to the existing $sup(X, \bigcup_{i=1}^{j-1} B_i)$.

This insertion (of new popular patterns) and update (of existing popular patterns) step is repeated for each batch.

Note that, during the mining process, our proposed Pop-streaming algorithm updates the $sumTL(X)$ and $sup(X)$ values stored in the Pop-stream structure whenever a batch of streaming data flows in. However, the algorithm does not repeatedly update $Pop(X)$. It uses the delay mode for mining: It only computes $Pop(X)$ based on the updated $sumTL(X)$ and $sup(X)$ values when the user needs the results. See Fig. 4 for a skeleton of the Pop-streaming algorithm.

Algorithm Pop-streaming

1. For each batch B_j do
   2. Apply Pop-growth to B_j to find popular patterns from B_j.
   3. Insert each mined popular pattern X into the Pop-stream structure and update its $sumTL(X)$ and $sup(X)$ values:
      3a. If the nodes corresponding to X do not exist in the Pop-stream structure, then create new nodes (each of which keeps $sumTL(X)$ and $sup(X)$ values) for X;
      3b. else, add its $sumTL(X)$ and $sup(X)$ values in B_j to the existing $sumTL(X)$ and $sup(X)$ values.
4. When a user requests the mining results (i.e., popular patterns) do
   5. Compute $Pop(X)$ based on the updated $sumTL(X)$ and $sup(X)$ values stored in the Pop-stream structure .
   6. If $Pop(X) \geq minpop$, then return X to the user.

Fig. 4. A skeleton of the Pop-streaming algorithm

7   Experimental Results

For experiments, we mostly use those datasets commonly used in frequent pattern mining experiments because characteristics of those transactional datasets are well known (see Table 4). More specially, we used (i) IBM synthetic datasets (e.g., T10I4D1M, T10I4D100K, T20I4D100K). and (ii) real datasets (e.g., chess, mushroom, connect-4) from the Frequent Itemset Mining Dataset Repository http://fimi.ua.ac.be/data/. We obtained consistent results for all of these datasets. Hence, to avoid repetition, we report here the experimental results on only a subset of these datasets in the remainder of this section.

Table 4. Dataset characteristics

Dataset	#transactions	#items	maxTL	avgTL	Data density
T10I4D100K	100,000	870	29	10.10	Sparse
T20I4D100K	99,996	871	42	19.81	Sparse
mushroom	8,124	119	23	23.00	Dense

All programs were written in C and run on UNIX with a quad-core processor with 1.3 GHz. The runtime specified indicates the total execution time (i.e., CPU and I/Os). The reported results are based on the average of multiple runs for each case. In all of the below experiments, Pop-trees were constructed using descending order of occurrence counts of items.

To the best of our knowledge, our Pop-tree is the first approach to mine popular patterns from transactional databases. Here, we first present the performance of our Pop-tree structure and Pop-growth algorithm when varying the mining parameters such as popularity threshold and dataset characteristics.

7.1   Runtime of Pop-Growth

In this section, we report the execution time that the Pop-growth requires for mining popular patterns over datasets of different types and changes in *minpop*. The execution time includes all the steps of H-table construction, the Pop-tree building and the corresponding mining. The results on one sparse dataset (e.g., T20I4D100K) and one dense dataset (e.g., mushroom) are presented in Fig. 5.

To observe the effect of mining on the variation in size of such datasets, we performed popular pattern mining while increasing the size of both of the

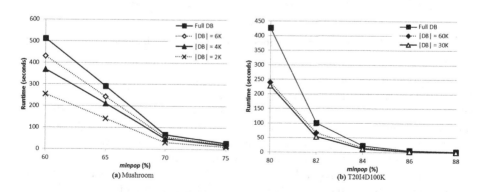

Fig. 5. Runtime of Pop-growth in mining transactional databases

datasets: (i) From 2 K to full for the mushroom dataset and (ii) from 30 K to full for T20I4D100K. Thus, the series for "Full DB" represent the results for the full size of datasets. Both datasets required more execution time when mining larger datasets. As the database size increased and *minpop* decreased, the tree structure size and number of popular patterns increased. Hence, a comparatively longer time was required to generate large numbers of popular patterns from large trees. Although the mushroom dataset is smaller in size, the transaction lengths of all transactions are the same (i.e., 23). Hence, the Pop-tree mining took a longer time when compared to a dataset with variable length such as T20I4D100K. The experimental results show that mining the corresponding Pop-tree for popular patterns is time efficient for both sparse and dense datasets.

7.2   Reduction on the Number of Patterns When Changing *minpop*

Similar to the previous experiment, we also examined the number of patterns generated by our Pop-growth when we varied the dataset size and *minpop*. Figure 6 shows the reduction in the number of patterns in percentage when increasing the *minpop* values in both the mushroom and T20I4D100K datasets with different dataset size. Each data point in the x-axes of the graphs reports the change of *minpop* from a low to a high value, while the y-axes indicate the percentage change in the number of patterns generated from a low to a high *minpop* value.

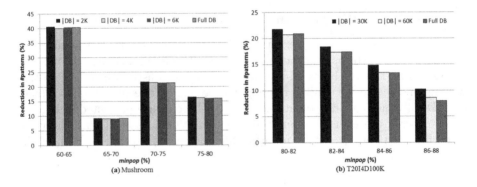

Fig. 6. Reduction on the number of patterns when changing *minpop*

Note that, depending on dataset characteristics, the reduction rate varied. For example, for the mushroom dataset, the reduction rate dropped sharply when *minpop* was changed from 60 %–65 % to 65 %–70 %, but the reduction rate rose when *minpop* was changed to 70 %–75 %. In contrast, T20I4D100K showed a consistent reduction rate when lowering the *minpop* value. However, as observed from the graphs for both datasets, the number of patterns reduced when increasing the *minpop* values. For example, for the mushroom dataset, the reduction rate was around 40 % when increasing the threshold from 60 % to 65 %. For 30 K of T20I4D100K, the reduction rate was around 21 % when increasing

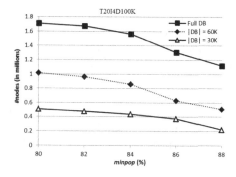

Fig. 7. Compactness of the Pop-tree: node count on T20I4D100K

the threshold from 80 % to 82 %. It is also interesting to note that the pattern count reduction rate was very similar irrespective of the database size.

We observed that the pattern generation characteristics of the proposed popular pattern mining algorithm were consistent with the variation of *minpop* and database size.

7.3   Compactness of the Pop-Tree

Here, we report the compactness of a Pop-tree in terms of number of Pop-tree nodes. Note that, as the mushroom dataset has a fixed transaction length, the maximum transaction length for every possible pattern in the dataset is always the same. Consequently, every item in the dataset passes the lazy pruning phase and contributes to the tree. Hence, for a particular portion of the mushroom dataset, the tree size (i.e., number of nodes) is the same with the variation of *minpop*. However, the number of nodes varied from 34523 (when $|\text{TDB}| = 2\,\text{K}$) to 91338 (when $|\text{TDB}| = 6\,\text{K}$). For the full dataset, it is around $100\,\text{K}$. The compactness of Pop-tree on different portion of T20I4D100K is presented in Fig. 7. The size of the tree structure gradually reduced in T20I4D100K with the increase of *minpop*.

As expected, in both datasets, the number of nodes increased with the increase in size of database. However, as far as the total number of nodes is concerned, one can observe that, irrespective of fixed or variable transaction length, a Pop-tree structure is compact enough to fit into a reasonable amount of memory.

7.4   Scalability of Pop-Growth

To study the scalability of Pop-growth mining technique, we further ran our algorithm on T10I4D100K, which is sparser than T20I4D100K. Figure 8 presents the results on scalability tests on the variation of *minpop* and required number of nodes on the dataset. Clearly, as the *minpop* decreases, the overall tree construction and mining time (Fig. 8(a)), and required memory (Fig. 8(b)) increase.

However, the Pop-tree shows a stable performance with a linear increase in run-time and memory consumption as the *minpop* decreased for the dataset. More-over, the results demonstrate that, the Pop-tree can mine the set of popular patterns on this dataset for a reasonably small value of popularity threshold with a considerable amount of execution time and memory.

To recap, the above experimental results show that the proposed Pop-tree can mine the set of popular patterns in both time and memory efficient manner over different types of dataset. Furthermore, the Pop-tree structure and Pop-growth algorithm are scalable for popularity threshold values and memory.

7.5   Mining Popular Friends from Social Networks

The aforementioned results show the time-efficiency of our proposed Pop-growth algorithm and the space-efficiency of our proposed Pop-tree structure for mining popular patterns from transactional data. Here, we experimented the efficiency of the Pop-growth algorithm when adapted to mine popular friends from social networks. To conduct this experiment, we used the social network datasets (e.g., Facebook, Twitter) from Stanford Large Network Dataset Collection https:// snap.stanford.edu/data/. For example, when *minpop* was set to 1043, Pop-growth only took 43 s to find about 384 K popular friend groups from the Face-book dataset (where $maxTL = 1045$). As another example, when *minpop* was set to 1203, Pop-growth took 70 s to find more (e.g., around 484 K) popular friend groups from another dataset—namely, the Twitter dataset (where $maxTL$ is higher and with a length of 1205).

7.6   Runtime of Pop-Streaming

After performing a series of experiments on popular pattern mining from static transactional databases or static social networks, we conduct experiments on popular pattern mining from dynamic data streams. Here, we divided datasets into multiple batches. We report the execution time that the Pop-streaming

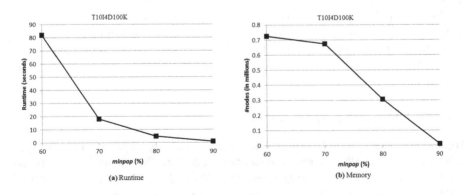

Fig. 8. Scalability on Pop-growth

requires for mining popular patterns over batches of streaming data of different types and changes in *minpop*. The execution time includes all the steps of the application of Pop-growth, the construction of the Pop-stream structure, and the corresponding mining. The results on dense streams (e.g., mushroom) and sparse streams (e.g., IBM) are presented in Fig. 9. Consistent with the runtime results on transactional database mining, Pop-streaming required shorter runtimes when *preMinpop* increased.

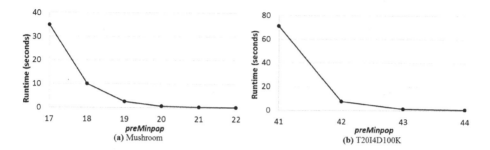

Fig. 9. Runtime of Pop-streaming in mining batches of data streams

7.7   Compactness of the Pop-Stream Structure

Next, we report the compactness of the Pop-stream structure in terms of number of nodes, which corresponds to the number of popular patterns mined from batches of streaming data. Figure 10 shows that, when *preMinpop* increased, the number of mined popular patterns decreased and thus reducing the size (i.e., reducing the number of nodes) of the Pop-stream structure.

7.8   Memory Consumption for the Pop-Streaming Algorithm

Recall that the first step of the Pop-streaming algorithm is to call Pop-growth for finding popular patterns from each batch of streaming data. When the Pop-growth algorithm is called, it builds a Pop-tree to capture important contents of transactions in the batch. When mining from w batches of the streaming data, the Pop-growth algorithm may be called w times. The size of the ing Pop-tree may vary from one batch to another batch. Figure 11 shows the maximum memory consumption among w Pop-trees.

Once the popular patterns are mined from a batch of streaming data, these mined patterns are then stored in the Pop-stream structure. Recall that, in Sect. 7.7, we measured the compactness of the Pop-stream structure. Note that memory consumption of the Pop-streaming algorithm mainly depends on that of the Pop-tree (measured in this section) and that of the Pop-stream structure (measured in Sect. 7.7). Hence, based on the experimental results from these two sections, we gained some insight about the amount of memory space required by the Pop-streaming mining process.

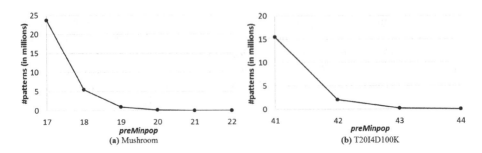

Fig. 10. Compactness of the Pop-stream structure: node count

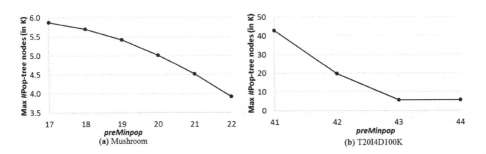

Fig. 11. Compactness of the Pop-tree structure in mining data streams: node count

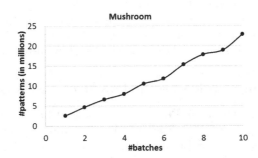

Fig. 12. Compactness of the Pop-stream structure when varying #batches

7.9   Scalability of the Pop-Stream Structure

Finally, we studied the scalability of the Pop-stream structure. In particular, Fig. 12 shows that, when the number of batches increased, the number of patterns to be stored in the Pop-stream structure gradually increased. Hence, the total number of nodes in the Pop-stream structure increased accordingly. The increase in the number of mined popular patterns (i.e., in the number of stored nodes) is proportional to the number of batches in the streaming data.

8   Conclusions and Future Work

In this paper, we introduced a new type of patterns—namely, *popular patterns*. We also proposed the *Pop-tree* (which captures important contents of transactional databases for mining popular patterns) and the *Pop-growth* algorithm (which finds popular patterns by mining the Pop-tree). Although the notion of popularity does not satisfy the downward closure property, we managed to address this issue by using total transaction length ($sumTL$) together with projected databases, which allows lazy pruning. Moreover, we also proposed the *Pop-stream* structure (which captures the popular patterns mined from each batch as well as other auxiliary information for computing the popularity of these patterns) and the *Pop-streaming* algorithm (which finds popular patterns by mining the Pop-stream. Moreover, results also showed that construction of Pop-tree and mining of popular patterns are time efficient. Furthermore, we are not confined with mining popular patterns from static transactional databases; we also mine popular patterns from dynamic data streams. Experimental results showed that both Pop-tree and Pop-stream structures are compact, scalable, and space efficient for both sparse and dense datasets (e.g., IBM synthetic data, real data from FIMI, social network data).

As future work, we plan to further extend our proposed framework as to incorporate novel extensions, precisely targeting several achievements: (i) incorporating the capability of dealing with *Big Data* (e.g., [8]), perhaps by adopting consolidated *data fragmentation approaches* (e.g., [3]), which well-adapt to massive sizes that both transactional databases and multi-rate, heterogeneous data streams may achieve; (ii) incorporating the capability of dealing with *optimization issues* (e.g. [5,6]), perhaps by adopting non-conventional approaches like *topology control* (e.g., [13]), which well-adapts to the graph-based nature of both connected entities in transactional databases and data stream items; (iii) incorporating the capability of dealing with *uncertain and imprecise* transactional databases (e.g., [27]) and data streams (e.g., [4]), perhaps by adopting probabilistic methods (e.g., [7]).

Acknowledgement. This project is partially supported by (i) China Scholarship Council (CSC), (ii) Mitacs (Canada), (iii) Natural Sciences and Engineering Research Council of Canada (NSERC), and (iv) University of Manitoba.

References

1. Agrawal, R., Srikant, R.: Fast algorithms for mining association rules. In: VLDB 1994, pp. 487–499 (1994)
2. Bailey, J., Manoukian, T., Ramamohanarao, K.: Fast algorithms for mining emerging patterns. In: Elomaa, T., Mannila, H., Toivonen, H. (eds.) PKDD 2002. LNCS (LNAI), vol. 2431, pp. 39–50. Springer, Heidelberg (2002)
3. Bonifati, A., Cuzzocrea, A.: Storing and retrieving XPath fragments in structured P2P networks. Data Knowl. Eng. **59**(2), 247–269 (2006)

4. Cuzzocrea, A.: Retrieving accurate estimates to OLAP queries over uncertain and imprecise multidimensional data streams. In: Cushing, J.B., French, J., Bowers, S. (eds.) SSDBM 2011. LNCS, vol. 6809, pp. 575–576. Springer, Heidelberg (2011)
5. Cuzzocrea, A., Furfaro, F., Greco, S., Masciari, E., Mazzeo, G.M., Saccà, D.: A distributed system for answering range queries on sensor network data. In: IEEE PerCom 2005 Workshops, pp. 369–373 (2005)
6. Cuzzocrea, A., Furfaro, F., Masciari, E., Saccà, D., Sirangelo, C.: A distributed system for answering range queries on sensor network data. In: Stefanidis, A., Nittel, S. (eds.) GeoSensor Networks, pp. 53–72. CRC Press (2004)
7. Cuzzocrea, A., Gunopulos, D.: A decomposition framework for computing and querying multidimensional OLAP data cubes over probabilistic relational data. Fundamenta Informaticae $132(2)$, 239–266 (2014)
8. Cuzzocrea, A., Saccà, D., Ullman, J.D.: Big data: a research agenda. In: IDEAS 2013, pp. 198–203. ACM (2013)
9. Cameron, J.J., Leung, C.K.-S., Tanbeer, S.K.: Finding strong groups of friends among friends in social networks. In: IEEE DASC 2011, pp. 824–831 (2011)
10. Cao, F., Ester, M., Qian, W., Zhou, A.: Density-based clustering over an evolving data stream with noise. In: SDM 2006, pp. 328–339. SIAM (2006)
11. Castellanos, M., Gupta, C., Wang, S., Dayal, U.: Leveraging web streams for contractual situational awareness in operational BI. In: EDBT/ICDT 2010 Workshops, art. 7. ACM (2010)
12. Chen, Y., Nascimento, M.A., Ooi, B.C., Tung, A.K.H.: SpADe: on shape-based pattern detection in streaming time series. In: IEEE ICDE 2007, pp. 786–795 (2007)
13. Cuzzocrea, A., Papadimitriou, A., Katsaros, D., Manolopoulos, Y.: Edge betweenness centrality: a novel algorithm for QoS-based topology control over wireless sensor networks. J. Netw. Comput. Appl. $35(4)$, 1210–1217 (2012)
14. Gaber, M.M., Zaslavsky, A.B., Krishnaswamy, S.: Mining data streams: a review. SIGMOD Rec. $34(2)$, 18–26 (2005)
15. Giannella, C., Han, J., Pei, J., Yan, X., Yu, P.S.: Mining frequent patterns in data streams at multiple time granularities. In: Kargupta, H., Joshi, A., Sivakumar, K., Yesha, Y. (eds.) Data Mining: Next Generation Challenges and Future Directions, pp. 105–124. AAAI/MIT Press (2004)
16. Gupta, A., Bhatnagar, V., Kumar, N.: Mining closed itemsets in data stream using formal concept analysis. In: Pedersen, T.B., Mohania, M.K., Tjoa, A.M. (eds.) DaWaK 2010. LNCS, vol. 6263, pp. 285–296. Springer, Heidelberg (2010)
17. Han, J., Pei, J., Yin, Y.: Mining frequent patterns without candidate generation. In: ACM SIGMOD 2000, pp. 1–12 (2000)
18. Jiang, N., Gruenwald, L.: Research issues in data stream association rule mining. SIGMOD Rec. $35(1)$, 14–19 (2006)
19. Lakshmanan, L.V.S., Leung, C.K.-S., Ng, R.T.: Efficient dynamic mining of constrained frequent sets. ACM Trans. Database Syst. $28(4)$, 337–389 (2003)
20. Lee, Y.-K., Kim, W.-Y., Cai, Y.D., Han, J.: CoMine: efficient mining of correlated patterns. In: IEEE ICDM 2003, pp. 581–584 (2003)
21. Leung, C.K.-S., Cuzzocrea, A., Jiang, F.: Discovering frequent patterns from uncertain data streams with time-fading and landmark models. T. Large-Scale Data- and Knowl.-Centered Syst. $\mathbf{8}$, 174–196 (2013)
22. Leung, C.K.-S., Hao, B.: Mining of frequent itemsets from streams of uncertain data. In: IEEE ICDE 2009, pp. 1663–1670 (2009)
23. Leung, C.K.-S., Jiang, F.: Frequent pattern mining from time-fading streams of uncertain data. In: Cuzzocrea, A., Dayal, U. (eds.) DaWaK 2011. LNCS, vol. 6862, pp. 252–264. Springer, Heidelberg (2011)

24. Leung, C.K.-S., Sun, L.: A new class of constraints for constrained frequent pattern mining. In: ACM SAC 2012, pp. 199–204 (2012)
25. Leung, C.K.-S., Tanbeer, S.K.: Mining popular patterns from transactional databases. In: Cuzzocrea, A., Dayal, U. (eds.) DaWaK 2012. LNCS, vol. 7448, pp. 291–302. Springer, Heidelberg (2012)
26. Leung, C.K.-S., Tanbeer, S.K.: Mining social networks for significant friend groups. In: Yu, H., Yu, G., Hsu, W., Moon, Y.-S., Unland, R., Yoo, J. (eds.) DASFAA Workshops 2012. LNCS, vol. 7240, pp. 180–192. Springer, Heidelberg (2012)
27. Motro, A.: Imprecision and uncertainty in database systems. In: Base, P., Kacprzyk, J. (eds.) Fuzziness in Database Management Systems. pp. 3–22. Physica-Verlag (1995)
28. Ng, W., Dash, M.: Discovery of frequent patterns in transactional data streams. T. Large-Scale Data- and Knowl.-Centered Syst. 2, 1–30 (2010)
29. Rasheed, F., Alshalalfa, M., Alhajj, R.: Efficient periodicity mining in time series databases using suffix trees. IEEE Trans. Knowl. Data Eng. 23(1), 79–94 (2011)
30. Rashid, M.M., Karim, M.R., Jeong, B.-S., Choi, H.-J.: Efficient mining regularly frequent patterns in transactional databases. In: Lee, S., Peng, Z., Zhou, X., Moon, Y.-S., Unland, R., Yoo, J. (eds.) DASFAA 2012, Part I. LNCS, vol. 7238, pp. 258–271. Springer, Heidelberg (2012)
31. Wu, T., Chen, Y., Han, J.: Re-examination of interestingness measures in pattern mining: a unified framework. Data Min. Knowl. Disc. 21(3), 371–397 (2010)
32. Xiong, H., Tan, P.-N., Kumar, V.: Hyperclique pattern discovery. Data Min. Knowl. Disc. 13(2), 219–242 (2006)
33. Yao, H., Hamilton, H.J.: Mining itemset utilities from transaction databases. Data Knowl. Eng. 59(3), 603–626 (2006)
34. Zhang, M., Kao, B., Cheung, D.W., Yip, K.Y.: Mining periodic patterns with gaprequirement from sequences, ACM Trans. Knowl. Discov. Data 1(2), art. 7 (2007)

Rare Pattern Mining from Data Streams Using SRP-Tree and Its Variants

David Tse Jung Huang$^{(\boxtimes)}$, Yun Sing Koh, and Gillian Dobbie

Department of Computer Science, University of Auckland,
Auckland, New Zealand
{dtjh,ykoh,gill}@cs.auckland.ac.nz

Abstract. There has been some research in the area of rare pattern mining where the researchers try to capture patterns involving events that are unusual in a dataset. These patterns are considered more useful than frequent patterns in some domains, including detection of computer attacks, or fraudulent credit transactions. Until now, most of the research in this area concentrates only on finding rare rules in a static dataset. There is a proliferation of applications which generate data streams, such as network logs and banking transactions, and applying techniques that mine static datasets is not practical for data streams. We propose a novel approach called Streaming Rare Pattern Tree (SRP-Tree) and its variations, which finds rare rules in a data stream environment using a sliding window, and show that it both finds the complete set of itemsets and runs with fast execution time.

Keywords: Rare pattern mining · FP-Growth · Data stream · Sliding window

1 Introduction

Traditionally pattern mining techniques focus on finding frequent patterns within a dataset. The early works in pattern mining revolve around the Apriori-like candidate generation-and-test approach. The Apriori algorithm [2] was widely used and studied with several variations proposed [1,15,22]. The introduction of FP-Tree by Han et al. [12] directed focus in frequent pattern mining from Apriori approaches to tree-structured approaches. Along with the introduction of FP-Tree was the simultaneous proposal of FP-Growth in [12]. Since then FP-Growth have been one of the most widely used tree mining algorithms in the recent decade.

Even though frequent patterns are widely considered to be both informative and useful, in some scenarios rare patterns may be more interesting as they represent infrequent events. Rare patterns are patterns that do not occur frequently within the dataset and can be considered as exceptions. The discovery of rare patterns also has possible application in several different areas of research such as Wireless Sensor Networks [8] and Auction Fraud Detection [30]. An example of a useful rare pattern in practice could be the association of certain occurrences of symptoms to diseases. For instance, Meningitis is the inflammation of the

© Springer-Verlag Berlin Heidelberg 2015
A. Hameurlain et al. (Eds.): TLDKS XXI, LNCS 9260, pp. 140–160, 2015.
DOI: 10.1007/978-3-662-47804-2_7

protective membranes covering the brain and spinal cord. Symptoms of Meningitis include headache, fever, vomiting, neck stiffness, and altered consciousness. Most of the symptoms commonly occur with influenza except for neck stiffness. By discovering the rare occurrence of all symptoms including neck stiffness, we are capable of flagging a patient possibly suffering from Meningitis out of a pool of patients suffering from common influenza. In recent years, the problem of extracting rare patterns from static datasets has been addressed. These proposed algorithms often follow either the Apriori algorithm or the FP-Tree algorithm. However, as the capability of generating data streams increases, the ability to capture useful information from streaming data becomes more important.

Ever since its inception, data stream mining has remained one of the more challenging problems within the data mining discipline. This is mainly due to the nature of data streams being continuous and unbounded in size as opposed to traditional databases where data is static and stable. Therefore, techniques developed have focused on improving the execution time and storage efficiency [7]. Recently the term "Big Data" has been widely used and has close relations to high performance data mining. Some of the important aspects of Big Data is introduced in [9] and the problem of devising models and algorithms for such high performance data mining tasks is further explained in [6].

Data from a wide variety of application areas ranging from online retail applications such as online auctions and online bookstores, telecommunications call data, credit card transactions, sensor data, wireless sensor networks and climate data are a few examples of applications that generate vast quantities of data on a continuous basis. Data produced by such applications are highly volatile with new patterns and trends emerging on a continuous basis. The unbounded size of data streams is considered the main obstacle when processing data streams. As it is unbounded, it makes it infeasible to store the entire data on disk. Furthermore, the processing of data streams should ideally be near real time. This raises two issues. Firstly, a multi-pass algorithm cannot be used because the entire dataset would need to be stored before mining can commence. This restriction limits the use of the majority of current database-based techniques as most of them require multiple scans of the entire dataset which needs the entire dataset to be stored. Secondly, obtaining the exact set of rules that includes both frequent and rare rules from the data streams is too expensive. In theory, most frequent pattern mining techniques that uses a minimum support threshold can be used to find rare patterns as these thresholds can be lowered to find both frequent and rare rules then later the frequent rules can be pruned out leaving only rare rules. However, this approach is extremely inefficient as a large number of unnecessary rules are generated and then discarded. In light of this situation, the need for an efficient algorithm that finds only rare rules in data streams without having to generate frequent rules is evident. In this paper we propose a technique that achieves this aim.

Contributions. Our contribution in this paper is the proposal of a novel technique called Streaming Rare Pattern Tree (SRP-Tree), which captures the complete set of rare rules in data streams using only a single pass scanning of the dataset and adapts a sliding window approach. We also propose two variations of the technique where the first one involves using a novel data structure called the Connection Table which allows us to efficiently constrain the search in our tree, and keep track of items within the window. The second variation improves upon the first and requires the tree to go through restructuring before mining with FP-Growth.

Paper Structure. The paper is organized as follows. In Sect. 2 we look at related work in the area of rare association rule mining. In Sect. 3 we present preliminary concepts and definitions for rare pattern mining in data streams. In Sect. 4 we describe our SRP-Tree approach, and in Sect. 5 we describe and discuss our experimental results. Finally, Sect. 6 concludes the paper.

2   Related Work

In this section we will discuss the related work in the area of pattern mining, specifically, frequent pattern mining and rare pattern mining. We look at the previous approaches that performs these tasks in both traditional static databases and in stream environments.

2.1   Frequent Pattern Mining

In the area of frequent pattern mining, the first algorithm was Apriori [2] and was proposed in the year 1994. The Apriori algorithm does multiple database scans and generates a large number of possible rules which are later found to be infrequent. This repeated candidate generation-and-test approach was extensively studied and used until in the year 2000 when the FP-Tree algorithm was proposed. The FP-Tree [12] is one of the most widely known techniques, and unlike Apriori it uses a tree structure to find frequent patterns in a database. FP-Tree solves the efficiency problem of Apriori-like approaches when finding patterns and rules and proposes the FP-Growth algorithm as the mining algorithm for the tree. FP-Tree proves to be efficient at finding the complete set of frequent patterns from a database and the FP-Growth algorithm was later adopted by many and considered as an efficient pattern mining algorithm. FP-Tree, being a landmark technique, is designed as a multi-pass approach which is not ideal in the data stream environment where data can only be read once. In addition, it is also memory intensive and unfitting for stream environments as streams are considered to be unbounded in data and working algorithms are required to have an efficient memory use.

Recently, more research that considers mining frequent patterns from data streams have been proposed. Giannella et al. [11] proposed a technique called FP-Stream which adopts the structure of FP-Tree and finds the approximate set of frequent patterns from data streams. FP-Stream mines time-sensitive frequent patterns and incrementally maintains only the historical information of frequent patterns. Overall, FP-Stream proposed efficient algorithms for constructing, maintaining, and updating pattern structures over data streams. However, FP-Stream only finds the approximate set of frequent patterns and discovering an approximate set of patterns is a disadvantage of the technique as in most cases, the user would want to find the complete set of patterns.

The technique DSTree was proposed by Leung et al. [20] and it is also a tree-structured algorithm. Unlike FP-Stream, DSTree finds the complete set of frequent patterns from a data stream. Unlike FP-Tree which builds its tree based on frequency-descending order, DSTree incrementally builds the tree based on a canonical ordering. DSTree is unaffected by changes in item frequency and therefore has the attractive property that it is easily updated and maintained as data is removed and added into the tree. DSTree has a wider applicability and can be used to find maximal, closed, and constrained itemsets.

Tanbeer et al. [27] proposed the CPS-Tree which is an efficient tree-structured algorithm that also finds the complete set of frequent patterns from a data stream. Unlike DSTree which uses a canonical ordering, CPS-Tree builds a compact tree in frequency-descending order by performing multiple sorting operations on the tree before mining of the tree. Resorting of the tree was proven to be more cost effective in the experiments when compared to other tree-based frequent mining techniques.

Most pattern mining algorithms like the DSTree and CPS-Tree adopt a sliding window model. However, for some applications, other models such the time-fading model and the landmark model may be more appropriate. In [19] the authors proposed a tree-based mining algorithm that mines frequent patterns from data streams using the time-fading model and the landmark model. The two different models were applied to streams of uncertain data were it is more sensible than using the sliding window model and the proposed algorithms were shown to be efficient through evaluation.

There has been a lot of other works in the area of data stream mining and most of them look at finding frequent patterns from data streams [3–5,13,16,18, 21,24].

2.2 Rare Pattern Mining

There has also been a lot of work in the area of rare pattern mining, including many recent works such as [17,23,25]. However all current research in this area is designed for static datasets and is not able to handle a data stream environment. Currently there are two different types of rare pattern mining approaches: level-wise and tree based, like that of frequent pattern mining. Current itemset mining approaches, which are based on level-wise exploration of the search space are similar to the Apriori algorithm [2]. In Apriori, k-itemsets (itemsets of cardinality k) are used to generate $k+1$-itemsets. These new $k+1$-itemsets are pruned using the downward closure property, which states that the superset of a non-frequent itemset cannot be frequent. Apriori terminates when there are no new $k+1$-itemsets remaining after pruning. MS-Apriori [22], Rarity [28], ARIMA [26], AfRIM [1] and Apriori-Inverse [15] are five algorithms that detect rare itemsets. They all use level-wise exploration similar to Apriori, which have candidate generation and pruning steps.

MS-Apriori [22] uses a bottom-up approach similar to Apriori. In MS-Apriori, each item can be assigned a different minimum item support value (MIS). Rare items can be assigned a low MIS, so that during candidate pruning, itemsets that include rare items are more likely to be retained and participate in rule generation. Apriori-Inverse [15] is used to mine perfectly rare itemsets, which are itemsets that only consist of items below a maximum support threshold (maxSup).

Szathmary et al. [26] proposed two algorithms that can be used together to mine rare itemsets: MRG-Exp and ARIMA. They defined three types of itemsets: minimal generators (MG), which are itemsets with a lower support than its subsets; minimal rare generators (MRG), which are itemsets with non-zero support and whose subsets are all frequent; and minimal zero generators (MZG), which are itemsets with zero support and whose subsets all have non-zero support. The first algorithm, MRG-Exp, finds all MRG by using MGs for candidate generation in each layer in a bottom up fashion. The MRGs represent a border that separates the frequent and rare itemsets in the search space. All itemsets above this border must be rare according to the antimonotonic property. The second algorithm, ARIMA, uses these MRGs to generate the complete set of rare itemsets. ARIMA stops the search for non-zero rare itemsets when the MZG border is reached, which represents the border above which there are only zero rare itemsets.

Adda et al. [1] proposed AfRIM, which begins with the itemset that contains all items found in the database. Candidate generation occurs by finding common k-itemset subsets between all combinations of rare $k + 1$-itemset pairs in the previous level. Troiano et al. proposed the Rarity algorithm that begins by identifying the longest transaction within the database and uses it to perform a top-down search for rare itemsets, thereby avoiding the lower layers that contain only frequent itemsets.

All of the above algorithms use the fundamental generate-and-test approach used in Apriori, which has potentially expensive candidate generation and pruning steps. In addition, these algorithms attempt to identify all possible rare itemsets, and as a result require a significant amount of execution time. RP-Tree algorithm was proposed by Tsang et al. [29] as a solution to these issues. RP-Tree avoids the expensive itemset generation and pruning steps by using a tree data structure, based on FP-Tree [12], to find rare patterns. However it uses a multi-pass approach, which is not suitable in a data stream environment.

More recently Lavergn et al. [17] proposed a rare pattern mining technique using itemset trees and relative support called TRARM-RelSup. As opposed to traditional techniques that used the minimum support threshold, TRARM-RelSup used the relative support measure with the goal of capturing some additional rare rules that are missed out when using the minimum support threshold. The algorithm combines the efficiency of targeted association mining querying with the capabilities of rare rule mining resulting in the discovery of more rare rules for the user.

Even though there has been many algorithms that perform rare pattern mining, up until now there has been no research into rare pattern mining in data streams.

3   Preliminaries

In this section, we provide definitions of key terms that explain the concepts of frequent pattern mining in a data stream. Let $\mathcal{I} = \{i_1, i_2, \ldots, i_n\}$ be a set of literals, called items, that represent a unit of information in an application domain. A set $X = \{i_l, \ldots, i_m\} \subseteq I$ and $l, m \in [1, n]$, is called a itemset, or a k-itemset if it contains k items. A transaction $t = (tid, Y)$ is a tuple where tid is a transaction-id and Y is a pattern. If $X \subseteq Y$, it is said that t contains X or X occurs in t. Let size(t) be the size of t, i.e., the number of items in Y.

Definition 1. A data stream $\mathcal{DS}$ is an infinite sequence of transactions $\mathcal{DS} = [t_1, t_2, \ldots, t_m)$, where $t_i, i \in [1, m]$ is the i th transaction in the data stream.

Definition 2. A window $\mathcal{W}$ is a set of all transactions between the ith and jth (where $j > i$) transactions and the size of $\mathcal{W}$ is $|\mathcal{W}| = j - i$.

Definition 3. Window $\mathcal{W}$ consists of blocks where $\mathcal{W} = \{B_1, B_2, \cdots, B_n\}$. A block is also a set of transactions like $\mathcal{W}$.

Definition 4. The count of an itemset X in $\mathcal{W}$, $count_\mathcal{W}(X)$, is the number of transactions in $\mathcal{W}$ that contain X.

Definition 5. The support of an itemset X in $\mathcal{W}$ is the count of an itemset divided by the size of $\mathcal{W}$

$$sup_\mathcal{W}(X) = \frac{count_\mathcal{W}(X)}{|\mathcal{W}|}$$

An association rule is an implication $X \to Y$ such that $X \cup Y \subseteq \mathcal{I}$ and $X \cap Y = \emptyset$. X is the antecedent and Y is the consequent of the rule. The *support* of $X \to Y$ in $\mathcal{W}$ is the proportion of transactions in $\mathcal{W}$ that contains $X \cup Y$. The *confidence* of $X \to Y$ is the proportion of transactions in $\mathcal{W}$ containing X that also contains Y.

3.1   Rare Itemsets

We adopted the rare itemsets concept from Tsang et al. [29]. We consider an itemset to be rare when its support falls below a threshold, called the minimum frequent support (minFreqSup) threshold. One difficulty when generating rare itemsets is differentiating noisy itemsets from the actual rare itemsets. As the support of the itemset is low, the potential of pushing unrelated items together increases as well, thus producing noisy itemsets. To overcome this problem, we define a noise filter threshold to prune out the noise called the minimum rare support (minRareSup) threshold. Typically minRareSup is set to a very low level, e.g., 0.01 %.

Definition 6. An itemset X is a *rare itemset* in a window $\mathcal{W}$ iff $sup_\mathcal{W}(X) \leq$ minFreqSup and $sup_\mathcal{W}(X) >$ minRareSup.

However not all rare itemsets that fulfill these properties are interesting. Furthermore, rare itemsets can be divided into two types: *rare-item itemsets* and *non-rare item itemsets*.

Rare item itemsets refer to itemsets which are a combination of only rare items and itemsets that consist of both rare and frequent items. Given 4 items $\{a, b, c, x\}$ with supports $a = 0.70$, $b = 0.45$, $c = 0.50$, and $x = 0.10$, with minFreqSup $= 0.15$ and minRareSup $= 0.01$, the itemset $\{a, x\}$ would be a rare item itemset assuming that the support of $\{a, x\} > 0.01$, since the itemset includes the rare item x.

Definition 7. An itemset X is a *rare-item itemset* iff X is a *rare itemset* **and**

$$\exists x \in X, sup_{\mathcal{W}}(x) \leq \text{minFreqSup}$$

Non-rare item itemsets only has frequent items which fall below the minimum frequent support threshold. Given 4 items $\{a, b, c, x\}$ with supports $a = 0.70$, $b = 0.45$, $c = 0.50$, and $x = 0.10$, with minFreqSup $= 0.15$ and minRareSup $= 0.01$, and the itemset $\{a, b, c\}$ with a support of 0.09, then this itemset would be a non-rare item itemset as all items within the itemset are frequent, and its support is between minFreqSup and minRareSup.

Definition 8. An itemset X is a *non-rare item itemset* iff X is a *rare itemset* **and**

$$\forall x \in X, sup_{\mathcal{W}}(x) > \text{minFreqSup}$$

4   SRP-Tree: Rare Pattern Tree Mining for Data Streams

Current tree based rare pattern mining approaches follow the traditional FP-Tree [12] approach. It is a two-pass approach and is affordable when mining a static dataset. However in a data stream environment, a two-pass approach is not suitable. To process a static environment, non-streaming rare pattern techniques such as RP-Tree order each item within a transaction according to its frequency before inserting it into the tree. Using these algorithms the frequency of the items are obtained during the first pass through the dataset.

In data streams we can only look at the transactions within the stream once, thus, a one-pass approach is necessary. This rules out the possibility of building a tree based on the frequency of items within the data stream. Furthermore, frequency of an item may change as the stream progresses. There are four scenarios which we need to consider:

Scenario 1. A frequent item x at a particular time T_1 may become rare at time T_2.

Scenario 2. A rare item x at a particular time T_1 may become frequent at time T_2.

Scenario 3. A frequent item x at a particular time T_1 may remain frequent at time T_2.

Scenario 4. A rare item x at a particular time T_1 may remain rare at time T_2.

T_1 represents a point in time, and T_2 represents a future point in time after T_1.

To find rare patterns within data streams, we propose a new algorithm called SRP-Tree that mines rare patterns in a data stream using a sliding window and tree based approach. We discuss the details of the algorithm and introduce the two variations of the algorithm: SRP-Tree with Connection Table and SRP-Tree with Restructuring.

4.1   Approach 1: SRP-Tree with Connection Table

In this approach, items from incoming transactions in the stream are inserted into a tree based on a canonical ordering. A canonical ordering allows us to capture the content of the transactions from the data stream and organise tree nodes according to a particular order. We use the lexicographic ordering of the items as the canonical ordering here. When we use a canonical order to build the tree, the ordering of items is unaffected by the changes in frequency caused by incremental updates. There has been other work carried out using a canonical ordering to build trees for a data stream mining environment [10,20], but these research has been tailored to find frequent patterns.

In our tree built using canonical ordering, the frequency of a node in the tree is at least as high as the sum of frequencies of its children. However, this does not guarantee the downward closure property which exists in a tree ordered in frequency-descending order. The downward closure property in a traditional rare pattern tree mining algorithm, whereby, *rare-items will never be the ancestor of a non-rare item in the initial tree due to the tree construction process* is violated. Hence, we propose a novel item list called the Connection Table which keeps track of each unique item in the window and the items they co-occur with along with their respective frequencies.

The Connection Table used in this approach captures in the transactions only items that have a lower canonical ordering. For example, given a transaction in a canonical order of $\{a, b\}$ we store in the table that a is connected to b with a frequency of 1 but it does not store b is connected to a. This is because of the properties of the canonical ordering in the constructed tree: item a will always be the ancestor of item b. Since mining is carried out using a bottom-up approach, by mining item b, item a is also mined. The opposite does not hold since mining item a does not guarantee that item b is mined. Therefore, by using the Connection Table to keep track of connected items and adding them as arguments to FP-Growth in the mining phase, the complete set of rare-item itemsets can then be captured. The Connection table is designed using a hash map which allows for $O(1)$ access. In worst case scenarios, the table could reach size of $\frac{x(x+1)}{2}$ where x is the total number of items; however, in reality this is highly unlikely.

At any point of time should we decide to mine the current window, the initial tree of the current window is used to construct conditional pattern bases and conditional trees for each rare-item and their connected items in the Connection Table. We trigger the mining step when a block is filled. In this example the block size will be equal to the window size of 12. Note that only connection items with an occurring frequency greater than or equal to the minRareSup are included. Each conditional tree and corresponding item are then used as arguments for FP-Growth. The threshold used to prune items from the conditional trees is minRareSup. The union of the results from each of these calls to FP-Growth is a set of itemsets that contains a rare-item, or rare item itemsets.

Table 1. Set of transactions in a given window $\mathcal{W}$ of size 12

Tid	Transaction	Tid	Transaction	Tid	Transaction
1	{a, g, h}	5	{c, f, h}	9	{c, d, h}
2	{a, g, h, i}	6	{a, e, g, h}	10	{b, f}
3	{b, c, d, f}	7	{g}	11	{a, h}
4	{b, d, j}	8	{h}	12	{a}

Table 2. Connection table using the window of transactions listed in Table 1

Item	Items Co-occurred	Item	Items Co-occurred
a	{(e:1), (g:3), (h:4), (i:1)}	f	{(h:1)}
b	{(c:1), (d:2), (f:2), (j:1)}	g	{(h:1), (i:1)}
c	{(d:2), (f:2), (h:2)}	h	{(i:1)}
d	{(f:1), (h:1), (j:1)}	i	{∅}
e	{(g:1), (h:1)}	j	{∅}

Example. Given the dataset in Table 1, we show how the Connection Table is built in Table 2. The left column in Table 2 list the unique items in the window, whereas the right column lists the set of co-occurring items along with the co-occurrence frequency of that particular item to the item in the right column. For example, item c co-occurs twice with items d, f, and h.

SRP-Tree with Connection Table Algorithm. Our SRP-Tree with connection table algorithm is shown in Algorithm 1. This approach essentially performs in one pass the counting of item frequencies and the building of the initial tree. Therefore, in a given window $\mathcal{W}$, for each incoming transaction t, SRP-Tree first updates the list of item frequencies according to the transactions contained in the current window. We refer to this as updateItemFreqList(t) method, where we increment the counts of items contained in the new transaction and decrement the counts of items contained in the oldest transcations to be discarded from the window. SRP-Tree then updates the tree structure according to the transactions contained in the current window in a similar fashion, which is referred to as updateTree(t) method in Algorithm 1. We mine the tree after a particular number of transactions also known as a block. We refer to a preset block size as

---

Algorithm 1. SRP-Tree (Connection Table)

---

1. **Input:** DS, $\mathcal{W}$, $\mathcal{B}$, $minRareSup, minFreqSup$;
2. **Output:** $results$ (Set of rare item itemsets);

3. **while** exist(DS) **do**
4. $t \leftarrow$ new incoming transaction from DS;
5.   currentBlockSize $\leftarrow$ currentBlockSize + 1;
6.   updateItemFreqList(t);
7.   updateConnectionTable(t);
8. $tree \leftarrow$ updateTree(t);
9.   **Mining at the end of block**
10.   **if** $\mathcal{B} ==$ currentBlockSize **then**
11.   currentBlockSize $\leftarrow$ 0;
12. $results = \emptyset$;
13. $\mathcal{I} \leftarrow$ {all unique items in $\mathcal{W}$};
14. $\mathcal{R} \leftarrow \{i \in \mathcal{I} \mid sup_{\mathcal{W}}(i) \geq minRareSup \wedge sup_{\mathcal{W}}(i) < minFreqSup\}$;
15. $\mathcal{C} \leftarrow \{k \in \mathcal{R}, j \in connectionTable(k) \mid sup_{\mathcal{W}}(k) \geq minRareSup\}$;
16.   **for** item a in $tree$ **do**
17.   **if** $a \in \mathcal{R}$ or $a \in \mathcal{C}$ **then**
18.   construct a's conditional pattern-base and then a's conditional FP-Tree $Tree_a$;
19. $results \leftarrow results \cup$ FP-Growth($Tree_a, a$);
20.   **end if**
21.   **end for**
22.   **end if**
23. **end while**
24. **return** $results$;

$\mathcal{B}$. In this algorithm, $\mathcal{R}$ refers to the set of rare items and $\mathcal{C}$ refers to the set of items that co-occur with a particular rare item.

We would also like to point out, that another difference between SRP-Tree and a static rare pattern mining approach is that in SRP-Tree, the tree is built using all the transactions in the window, whereas in a static rare pattern mining approach, only transactions with rare items are used to build the tree. A static approach has the luxury of looking at the dataset twice and discarding items which it is not interested in before the tree is even built. In our case, we simply cannot know which transactions contain rare items until we decide to mine.

SRP-Tree with Connection Table Example. Applying SRP-Tree to the window $\mathcal{W}$ of the 12 transactions in Table 1, the support ordered list of all items is $\langle (h{:}6), (a{:}5), (g{:}4), (b{:}3), (c{:}3), (d{:}3), (f{:}2), (e{:}1), (i{:}1), (j{:}1) \rangle$. Using minFreqSup = 4 and minRareSup = 2, only the items $\{b, c, d, f\}$ are rare, and included in the set of rare items, $\mathcal{R}$.

The initial SRP-Tree constructed for window $\mathcal{W}$ of size 12 is shown in Fig. 1. To find the rare-item itemsets, the initial SRP-Tree is used to build conditional pattern bases and conditional SRP-Trees for each rare item $\{b, c, d, f\}$ and any additional items in the Connection Table that is connected with a rare item that

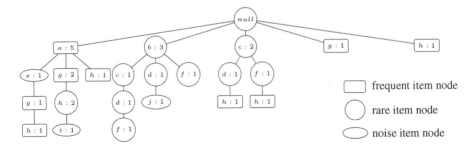

Fig. 1. Pattern tree constructed from window $\mathcal{W}$ using SRP-Tree

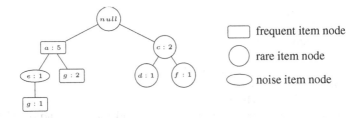

Fig. 2. Conditional tree, $Tree_h$

has a frequency greater than the minRareSup, in this example, item h. The conditional tree for item h is shown in Fig. 2. Each of the conditional SRP-Trees and the conditional item are then used as parameters for the FP-Growth algorithm.

4.2   Approach 2: SRP-Tree with Restructuring

In the previous approach we build the tree based on a canonical ordering and proposed the Connection Table that enables the capture of the complete set of rare rules from the canonically ordered tree. In this approach we improve upon the first and use an additional tree restructuring step. The additional restructuring step replaces the Connection Table and still allows the algorithm to find the complete set of rare rules.

The Connection Table proposed in the previous approach was used because when building the tree with a canonical ordering, the downward closure property is violated (see Sect. 4.1). In this second approach, instead of using the Connection Table, we perform a tree restructuring at the end of each block before mining the tree with FP-Growth. The additional tree restructuring effectively re-orders the nodes in the tree at each block into a strict frequency-descending order for that block. After restructuring, the tree will maintain the downward closure property and performing FP-Growth on the rare items themselves will yield the complete set of rare rules.

SRP-Tree with Restructuring Algorithm. The algorithm for the restructuring approach of our SRP-Tree is shown in Algorithm 2. The building and

Algorithm 2. SRP-Tree (Restructuring)

1. **Input:** DS, W, B, $minRareSup$, $minFreqSup$;
2. **Output:** $results$ (Set of rare item itemsets);

3. **while** exist(DS) **do**
4. $t \leftarrow$ new incoming transaction from DS;
5.   currentBlockSize $\leftarrow$ currentBlockSize $+ 1$;
6.   updateItemFreqList(t);
7. $tree \leftarrow$ updateTree(t);
8.   **Mining at the end of block**
9.   **if** $B ==$ currentBlockSize **then**
10.   currentBlockSize $\leftarrow 0$;
11. $tree \leftarrow$ restructureTree();
12. $results = \emptyset$;
13. $\mathcal{I} \leftarrow \{$all unique items in $W\}$;
14. $\mathcal{R} \leftarrow \{i \in \mathcal{I} \mid sup_W(i) \geq minRareSup \wedge sup_W(i) < minFreqSup\}$;
15.   **for** item a in $tree$ **do**
16.   **if** $a \in \mathcal{R}$ **then**
17.   construct a's conditional pattern-base and then a's conditional FP-Tree $Tree_a$;
18. $results \leftarrow results \cup$ FP-Growth($Tree_a, a$);
19.   **end if**
20.   **end for**
21.   **end if**
22. **end while**
23. **return** $results$;

updating of the item frequency list and the tree is similar to the previous approach and explained in the earlier sections. The main difference is the absence of maintaining the Connection Table and also an additional restructuring method at line 11.

Restructuring Technique. A tree is usually restructured by rearranging the nodes of an existing tree built based on a previous ordering to another different desired ordering. In our case we are restructuring the tree built based on a canonical ordering to a frequency-descending ordering. This operation involves sorting the item frequencies list and shifting the position of the nodes in the tree. We adopt the restructuring technique used in [16]. The branch sorting method (BSM) is an efficient tree restructuring technique that fits the purpose of our algorithm. In summary, BSM performs tree restructures by going through several steps. First the item-frequency-list of the items in the current block is rearranged into frequency-descending order. Then, the technique iterates through each unsorted path in the tree and sorts them based on the reordered item-frequency-list. BSM is an array-based technique whereby for each unsorted branch (path) in the tree, it first removes the path, sorts the path based on the frequency-descending ordering, then adds the path back into the tree.

Restructuring is complete when all the branches are processed and sorted which produces the final sorted tree.

SRP-Tree with Restructuring Example. Consider the same example as given by the transactions in Table 1. The support ordered list of all items is $\langle(h{:}6)$, $(a{:}5)$, $(g{:}4)$, $(b{:}3)$, $(c{:}3)$, $(d{:}3)$, $(f{:}2)$, $(e{:}1)$, $(i{:}1)$, $(j{:}1)\rangle$. Using minFreqSup = 4 and minRareSup = 2, only the items $\{b, c, d, f\}$ are rare, and included in the set of rare items, $\mathcal{R}$.

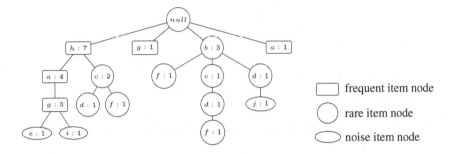

Fig. 3. Pattern tree from window $\mathcal{W}$ after restructuring into frequency descending order

In this approach, instead of mining the tree (shown in Fig. 1) built with canonical ordering $a, b, c, d, e, f, g, h, i, j$, we mine the restructured tree based on the support ordered list of items in the window W with the ordering $h, a, g, b, c, d, f, e, i, j$. The tree after restructuring is shown in Fig. 3.

Recall that the downward closure property in a traditional rare pattern tree mining algorithm states that *rare-items will never be the ancestor of a non-rare item in the initial tree due to the tree construction process*. From this restructured tree built based on frequency-descending support ordering, we can observe that all the rare-items (circular nodes) are indeed never the ancestor of a non-rare item (rectangular nodes). Therefore, we can prove that the downward closure property is satisfied in this case.

We also observe that because of the change in the ordering of the tree, it has become more compact (containing fewer nodes). In general, the compactness of a tree will have an influence on the mining time of FP-Growth. Therefore, the more compact the tree is, the faster the mining time of FP-Growth.

5   Experimental Results

In this section we present the evaluation and results performed on the two variations of our algorithm SRP-Tree. In Sect. 5.1 we describe in detail the algorithm we used to compare our SRP-Tree against. In Sect. 5.2 we evaluate the algorithms on real-world datasets and compare their performance at finding rare-item itemsets. Lastly in Sect. 5.3 we present a case study on the T40I10D100K dataset and perform additional evaluations by varying the minimum frequent support threshold and the block size of the techniques.

All algorithms were implemented in Java and executed on a machine equipped with an Intel Core i5-2400 CPU @ 3.10 GHz with 4 GB of RAM running Windows 7 x 64.

5.1   Streaming Canonical Tree (SC-Tree)

SRP-Tree is the very first attempt at mining rare patterns in a data stream environment so there are no other similar techniques that mine rare patterns in a data stream to compare to for evaluation. Therefore, in these experiments we compared the performance of our SRP-Tree to the Streaming Canonical Tree (SC-Tree), which is a modified version of DSTree [20] with pruning of frequent itemsets. This technique that we use for comparison, SC-Tree with pruning, finds rare item itemsets in an unoptimized brute-force manner. The SC-Tree is a one pass technique which stores the transactions from the stream in a tree using canonical ordering and stores/updates item frequencies similar to DSTree. To find all rare item itemsets, SC-Tree finds and generates all itemsets that meet the minRareSup threshold, then removes all other itemsets except rare item itemsets with an extra pruning step. SC-Tree produces the same itemsets and generates the same rules as SRP-Tree (Table) and SRP-Tree (Restructure).

5.2   Real-World Dataset

In this section we present the time and relative time taken for itemset generation of SRP-Tree (both approaches) and SC-Tree on real-world datasets. The time is reported in seconds and the relative time is calculated by setting SC-Tree to 1.00 and SRP-Tree relative to that of SC-Tree.

$$\text{relative time} = \frac{\text{Time taken by SRP-Tree}}{\text{Time taken by SC-Tree}}$$

We used a window size and block size of 25 K for all datasets except for the Mushroom dataset where we used 2 K due to its smaller size. The minFreqSup and minRareSup thresholds are shown in Table 3 for each dataset. The thresholds are user-defined through examining the distribution of item frequencies in each of the datasets. We acknowledge that given a data stream environment, this is not the most suitable way of defining thresholds and in the future we will be looking at a way to adapt the thresholds to the change in distribution and drift of the transactions in the stream.

We have tested the algorithms on 6 datasets obtained from the FIMI (Frequent Itemset Mining Implementations) repository[1]. The datasets are: Mushroom, Retail, BMS-POS, T10I4D100K, T40I10D100K, and Kosarak (250K).

Here are brief descriptions of each dataset:

---

[1] http://fimi.ua.ac.be/data/.

- The Mushroom dataset is a dense dataset with a relatively larger number of possible rules and patterns. It contains a total of 8124 instances and 22 attributes that includes descriptions of hypothetical samples corresponding to 23 species of gilled mushrooms in the Agaricus and Lepiota family.
- The Retail dataset, compared to Mushroom, is a much more sparse dataset that contains anonymized retail market basket data from an Belgian retail store over approximately 5 months. The total number of instances is 88163.
- The BMS-POS dataset contains several years worth of point-of-sale data from a large electronics retailer. Each transaction represents customer purchases of items at one time. It contains a total of 515597 instances and the average transaction size is 6.5.
- The T10I4D100K and T40I10D100K originates from the same source, the IBM Almaden Quest market basket data generator. They both contain a total of 100000 instances.
- The Kosarak dataset contains anonymized click-stream of a hungarian online news portal. Kosarak contains a larger number of different transactions with different characteristics (i.e. the dataset contains transactions that vary largely in average transaction size and items).

Table 3. Comparison between SRP-Tree and SC-Tree

Dataset	B	MinRareSup	MinFreqSup	# Itemsets	SRPT (Table)		SRPT (Restructure)		SC-Tree	
					Time (s)	Rel. time	Time (s)	Rel. time	Time (s)	Rel. time
Mushroom	2K	0.01	0.05	14443674	1131	0.86	86	0.07	1321	1.00
Retail	25K	0.0001	0.0005	572673	239	0.71	9	0.03	339	1.00
BMS-POS	25K	0.0002	0.0005	1426	58	0.02	25	0.01	3783	1.00
T10I4D100K	25K	0.0001	0.0005	1161	12	0.63	6	0.32	19	1.00
T40I10D100K	25K	0.003	0.05	4734806	301	0.79	94	0.25	380	1.00
Kosarak(250K)	25K	0.001	0.15	35623519	703	0.10	591	0.08	7213	1.00

Table 3 shows, for each dataset, the block size, minRareSup, and minFreqSup used to run the algorithms then the comparison between SRP-Tree (Table), SRP-Tree (Restructure) and SC-Tree of the itemsets generated, time it took to run the algorithm, and the relative time. The objective of this comparison is to look at the efficiency of each algorithm under the same conditions. SRP-Tree (Table), SRP-Tree (Restructure) and SC-Tree generate the same number of itemsets because both SRP-Tree approaches only generate rare-item itemsets and SC-Tree generates all itemsets that meet the minRareSup threshold then the extra pruning step removes all other itemsets that are not rare-item itemsets. Therefore, the final number of itemsets generated by both algorithms is the same.

In all datasets of varying item frequency distribution, both SRP-Tree approaches run faster than SC-Tree where SRP-Tree (Restructure) runs significantly faster than the other two techniques. The time taken to run the various datasets are highly dependent on the tree structure built from the transactions in the datasets and generally have a positive correlation with the number of itemsets generated. The number of itemsets generated also has a great variation and is highly dependent on the composition/nature of the respective dataset. For datasets that are more sparse in nature like Retail, BMS-POS, and T10I4D100K, the number of itemsets generated are usually smaller than a dense dataset like Mushroom and the run-time is usually faster.

5.3   Case Study: T40I10D100K

The T40I10D100K dataset is generated using the generator from the IBM Almaden Quest research group. Figures 4 and 5 shows the item frequency distribution of the T40I10D100K dataset. When we compared the distribution of the T40I10D100K dataset, to other datasets in the FIMI repository, we observed that the T40I10D100K dataset is more sparse in nature compared to the Mushroom dataset, but more dense than datasets like Retail and BMS-POS. It is also important to note that T40I10D100K contains a high proportion of items with a frequency of less than 0.1 (approximately 90 % of the total items). This particular distribution contains a greater number of prospective rare items and rules to be mined from the dataset.

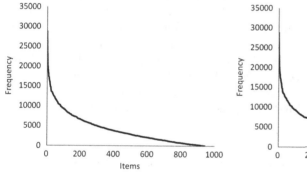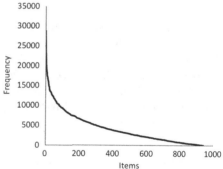

Fig. 4. Item frequency distribution

Fig. 5. Normalized item frequency distribution

In Table 4 we show the difference in itemsets generated and the time it took to generate items of varying minFreqSup on the T40I10D100K dataset. We aim to look at the fluctuations in run-time and the number of itemsets generated caused by varying the minimum frequent support threshold. As we increase the minFreqSup, the number of itemsets generated increases and the relative time also increases. It is important to note that the real time taken for SRP-Tree (Table) to generate itemsets decreases as we increase the minFreqSup. This is because at a lower minFreqSup for this particular distribution of the dataset, there is a high co-occurrence of rare items to other items with similar item frequency (indicating that these other items are also likely to have rare properties).

Table 4. Varying MinFreqSup for T40I10D100K

MinFreqSup	Itemset	SRPT (Table)		SRPT (Restructure)		SC-Tree	
		Time(s)	Relative Time	Time(s)	Relative Time	Time(s)	Relative Time
0.04	4397517	333	0.93	76	0.21	357	1.00
0.05	4734806	301	0.79	94	0.25	380	1.00
0.06	5028947	278	0.66	116	0.28	421	1.00
0.10	5105892	246	0.55	148	0.33	446	1.00
0.15	5136904	238	0.52	160	0.35	454	1.00

The mining of these additional highly connected items with rare association patterns incurred a larger overhead in maintaining the Connection Table. As the minFreqSup increases and most of the highly connected potential rare items are accounted for, the overhead decreases and this results in a faster runtime.

However, for SRP-Tree (Restructure) the time taken to generate itemsets increases as we increase the minFreqSup like that of SC-Tree. This is because this approach does not use the Connection Table structure to mine for itemsets and that the time taken is generally positively correlated with the number of itemsets. Table 5 shows the difference in execution time when the block size varies. Similar to the above experiment, we vary one variable in the algorithms while keeping the others constant. Here we aim to compare the behavior of the algorithms when the block size variable is changed. Overall we observed that the execution time is increased as the block size increases due to the increased size of the tree being built and SRP-Tree (Restructure) is still the fastest running technique of the three.

Table 5. Execution Time based on Varying Block Sizes for T40I10D100K

Block size	SRPT (Table)		SRPT (Restructure)		SC-Tree	
	Avg time / window (s)	Relative time	Avg. time / window (s)	Relative time	Avg. time / window (s)	Relative time
10K	41	0.59	19	0.27	70	1.00
25K	75	0.79	24	0.25	95	1.00
50K	142	0.71	19	0.09	201	1.00

6 Discussion

Rare pattern mining is often a more difficult and computationally expensive task than frequent pattern mining. Frequent patterns occur with a high frequency and frequent itemsets will only contain frequent items, therefore, frequent pattern mining techniques generally need to only consider a smaller subset of items out of the total set when mining. Strategies such as top-k items or minimum frequent support threshold are used to determine what items are frequent. The natural property of frequent items virtually constrains the search space of the mining techniques and effectively helps guarantee mining efficiency. In contrast, rare patterns occur with a low frequency and rare itemsets may contain both frequent and rare items. When mining rare patterns, the search space is considered to be much larger compared to mining frequent patterns, making rare pattern mining a more computationally expensive task. We remedy this by better constraining the initial search space of the patterns using SRP-Tree. We compared our two variations of SRP-Tree against SC-Tree, a modified version of DSTree that finds rare patterns using a brute-force strategy. Using SC-Tree, the search space included both frequent and rare patterns. It is only through an additional pruning step that SC-Tree removes frequent patterns from the final discovered set leaving the desired rare patterns. The effects of constraining the initial search space using SRP-Tree is evident from the experimental evaluation showing a significant improvement in execution time.

Traditionally pattern mining techniques are used in databases to find a specific set of useful patterns. In dynamically changing data streams, the set of patterns can vary and change over-time. Our SRP-Tree which works in a stream environment can be incorporated with drift detection methods [14] to provide an effective way of discovering useful rare patterns at various points of the data stream. Drift detection signals points where patterns have possibly changed, then instead of using a block size variable to mine patterns periodically, the drift points found by the detection method is used as points when SRP-Tree mines patterns. This incorporated approach will allow the user to discover more meaningful rare patterns and also during times when patterns do not change, saves on the unnecessary mining steps.

7   Conclusions and Future Work

We present a new algorithm for mining rare patterns using a tree structure in a data stream environment with two different variations. To the extent of our knowledge, this is the first algorithm that looks at mining rare patterns in a data stream. Our technique is a one-pass only strategy which is capable of mining rare patterns in a static database or in a dynamic data stream. In the case of mining data streams, our technique is also capable of mining at any given point of time in the stream and with different window and block sizes. One of the contributions of this algorithm is a novel approach using a Connection Table which keeps track of related items in a sliding window and reduces the mining space during itemset generation. In our evaluations on six different datasets, both variations of the SRP-Tree algorithm are capable of generating itemsets in a more efficient manner compared to the SC-Tree.

In the future we intend to investigate dynamically adapting the minRareSup and minFreqSup thresholds on-the-fly because data streams are volatile and neither setting fixed thresholds for all data nor defining the thresholds prior based on distribution is deemed suitable. We will also look at the possibility of dynamically adjusting the window size to reflect the density of incoming data in the stream. For example, if the new transactions in the window contained uninteresting or duplicate itemsets and rules, we could (through varying the window size) decide not to mine until more interesting itemsets and rules are captured. It will also be interesting to investigate the limitations of the tree with respect to the different characteristics and intensity of the data stream.

References

1. Adda, M., Wu, L., Feng, Y.: Rare itemset mining. In: Proceedings of the Sixth International Conference on Machine Learning and Applications, ICMLA 2007, pp. 73–80. IEEE Computer Society, Washington, DC (2007)
2. Agrawal, R., Srikant, R.: Fast algorithms for mining association rules in large databases. In: Bocca, J.B., Jarke, M., Zaniolo, C. (eds.) Proceedings of the 20th International Conference on Very Large Data Bases, VLDB, pp. 487–499. Morgan Kaufmann, Santiago (1994)
3. Cheng, J., Ke, Y., Ng, W.: Maintaining frequent closed itemsets over a sliding window. J. Intell. Inf. Syst. **31**, 191–215 (2008)
4. Chi, Y., Wang, H., Yu, P.S., Muntz, R.R.: Moment: maintaining closed frequent itemsets over a stream sliding window. In: Proceedings of the Fourth IEEE International Conference on Data Mining, ICDM 2004, pp. 59–66. IEEE Computer Society, Washington, DC (2004)
5. Chi, Y., Wang, H., Yu, P.S., Muntz, R.R.: Catch the moment: maintaining closed frequent itemsets over a data stream sliding window. Knowl. Inf. Syst. **10**, 265–294 (2006)
6. Cuzzocrea, A.: Models and algorithms for high-performance distributed data mining. J. Parallel Distrib. Computi. **73**(3), 281–283 (2013)
7. Cuzzocrea, A., Furfaro, F., Masciari, E., Saccà, D., Sirangelo, C.: Approximate query answering on sensor network data streams. In: GeoSensor Networks, vol. 49, pp. 53–72 (2004)
8. Cuzzocrea, A., Papadimitriou, A., Katsaros, D., Manolopoulos, Y.: Edge betweenness centrality: a novel algorithm for qos-based topology control over wireless sensor networks. J. Network Comput. Appl. **35**(4), 1210–1217 (2012). http://dx.doi.org/10.1016/j.jnca.2011.06.001
9. Cuzzocrea, A., Saccà, D., Ullman, J.D.: Big data: a research agenda. In: Proceedings of the 17th International Database Engineering & #38; Applications Symposium, IDEAS 2013, pp. 198–203. ACM, New York (2013)
10. Datar, M., Gionis, A., Indyk, P., Motwani, R.: Maintaining stream statistics over sliding windows: (extended abstract). In: Proceedings of theThirteenth Annual ACM-SIAM Symposium on Discrete Algorithms, SODA 2002, Society for Industrial and Applied Mathematics, Philadelphia, PA, USA, pp. 635–644 (2002)
11. Giannella, C., Han, J., Pei, J., Yan, X., Yu, P.S.: Mining frequent patterns in data streams at multiple time granularities (2002)
12. Han, J., Pei, J., Yin, Y.: Mining frequent patterns without candidate generation. In: Proceedings of the 2000 ACM SIGMOD International Conference on Management of Data, SIGMOD 2000, pp. 1–12. ACM, New York (2000)

13. Huang, D.T.J., Koh, Y.S., Dobbie, G., Pears, R.: Kernel-tree: mining frequent patterns in a data stream based on forecast support. In: Thielscher, M., Zhang, D. (eds.) AI 2012. LNCS, vol. 7691, pp. 614–625. Springer, Heidelberg (2012). http://dx.doi.org/10.1007/978-3-642-35101-3_52
14. Huang, D.T.J., Koh, Y.S., Dobbie, G., Pears, R.: Detecting changes in rare patterns from data streams. In: Tseng, V.S., Ho, T.B., Zhou, Z.-H., Chen, A.L.P., Kao, H.-Y. (eds.) PAKDD 2014, Part II. LNCS, vol. 8444, pp. 437–448. Springer, Heidelberg (2014). http://dx.doi.org/10.1007/978-3-319-06605-9_36
15. Koh, Y.S., Rountree, N.: Finding sporadic rules using apriori-inverse. In: Ho, T.-B., Cheung, D., Liu, H. (eds.) PAKDD 2005. LNCS (LNAI), vol. 3518, pp. 97–106. Springer, Heidelberg (2005)
16. Koh, Y.S., Dobbie, G.: Efficient single pass ordered incremental pattern mining. In: Hameurlain, A., Küng, J., Wagner, R., Cuzzocrea, A., Dayal, U. (eds.) TLDKS VIII. LNCS, vol. 7790, pp. 137–156. Springer, Heidelberg (2013). http://dx.doi.org/10.1007/978-3-642-37574-3_6
17. Lavergne, J., Benton, R., Raghavan, V.V.: TRARM-RelSup: targeted rare association rule mining using itemset trees and the relative support measure. In: Chen, L., Felfernig, A., Liu, J., Raś, Z.W. (eds.) ISMIS 2012. LNCS, vol. 7661, pp. 61–70. Springer, Heidelberg (2012). http://dx.doi.org/10.1007/978-3-642-34624-8_7
18. Lee, C.H., Lin, C.R., Chen, M.S.: Sliding window filtering: an efficient method for incremental mining on a time-variant database. Inf. Syst. 30(3), 227–244 (2005)
19. Leung, C.K.-S., Cuzzocrea, A., Jiang, F.: Discovering frequent patterns from uncertain data streams with time-fading and landmark models. In: Hameurlain, A., Küng, J., Wagner, R., Cuzzocrea, A., Dayal, U. (eds.) TLDKS VIII. LNCS, vol. 7790, pp. 174–196. Springer, Heidelberg (2013). http://dx.doi.org/10.1007/978-3-642-37574-3_8
20. Leung, C.K.S., Khan, Q.I.: Dstree: A tree-structure for the mining of frequent sets from data streams. In: Proceedings of the Sixth International Conference on Data Mining, ICDM 2006, pp. 928–932. IEEE Computer Society, Washington, DC (2006)
21. Li, H.F., Lee, S.Y.: Mining frequent itemsets over data streams using efficient window sliding techniques. Expert Syst. Appl. 36, 1466–1477 (2009)
22. Liu, B., Hsu, W., Ma, Y.: Mining association rules with multiple minimum supports. In: Proceedings of the 5th ACM SIGKDD International Conference on Knowledge Discovery and Data Mining, pp. 337–341 (1999)
23. Luna, J., Romero, J., Ventura, S.: On the adaptability of g3parm to the extraction of rare association rules. Knowl. Inf. Syst. 38, 391–418 (2013). http://dx.doi.org/10.1007/s10115-012-0591-9
24. Mozafari, B., Thakkar, H., Zaniolo, C.: Verifying and mining frequent patterns from large windows over data streams. In: Proceedings of the 2008 IEEE 24th International Conference on Data Engineering, pp. 179–188. IEEE Computer Society, Washington, DC (2008). http://dl.acm.org/citation.cfm?id=1546682.1547157
25. Okubo, Y., Haraguchi, M., Nakajima, T.: Finding rare patterns with weak correlation constraint. In: 2010 IEEE International Conference on Data Mining Workshops (ICDMW), pp. 822–829 (2010)
26. Szathmary, L., Napoli, A., Valtchev, P.: Towards rare itemset mining. In: Proceedings of the 19th IEEE International Conference on Tools with Artificial Intelligence, ICTAI 2007, vol. 01, pp. 305–312. IEEE Computer Society, Washington, DC (2007)
27. Tanbeer, S.K., Ahmed, C.F., Jeong, B.S., Lee, Y.K.: Sliding window-based frequent pattern mining over data streams. Inf. Sci. 179(22), 3843–3865 (2009)

28. Troiano, L., Scibelli, G., Birtolo, C.: A fast algorithm for mining rare itemsets. In: Proceedings of the 2009 Ninth International Conference on Intelligent Systems Design and Applications, ISDA 2009, pp. 1149–1155. IEEE Computer Society, Washington, DC (2009)

29. Tsang, S., Koh, Y.S., Dobbie, G.: RP-tree: rare pattern tree mining. In: Cuzzocrea, A., Dayal, U. (eds.) DaWaK 2011. LNCS, vol. 6862, pp. 277–288. Springer, Heidelberg (2011)

30. Tsang, S., Koh, Y.S., Dobbie, G., Alam, S.: SPAN: finding collaborative frauds in online auctions. Knowl.-Based Syst. **71**, 389–408 (2014). http://dx.doi.org/10.1016/j.knosys.2014.08.016

Improving Cross-Document Knowledge Discovery Through Content and Link Analysis of Wikipedia Knowledge

Peng Yan[✉] and Wei Jin

Department of Computer Science, North Dakota State University,
1340 Administration Ave., Fargo, ND 58102, USA
{peng.yan,wei.jin}@ndsu.edu

Abstract. The Vector Space Model (VSM) has been widely used in Natural Language Processing (NLP) for representing text documents as a Bag of Words (BOW). However, only document-level statistical information is recorded (e.g., document frequency, inverse document frequency) and word semantics cannot be captured. Improvement towards understanding the meaning of words in texts is a challenging task and sufficient background knowledge may need to be incorporated to provide a better semantic representation of texts. In this paper, we present a text mining model that can automatically discover semantic relationships between concepts across multiple documents (where the traditional search paradigm such as search engines cannot help much) and effectively integrate various evidences mined from Wikipedia knowledge. We propose this integration may effectively complement existing information contained in text corpus and facilitate the construction of a more comprehensive representation and retrieval framework. The experimental results demonstrate the search performance has been significantly enhanced against two competitive baselines.

Keywords: Knowledge discovery · Semantic relatedness · Cross-Document knowledge discovery · Document representation

1 Introduction

Text mining aims at mining high-quality information from mass text. However, great challenges have been posed for many text mining tasks because of the increasing sheer volume of text data and the difficulty of capturing valuable knowledge hidden in them. Therefore efficient and high-quality text mining algorithms are demanded and effective document representation and accurate semantic relatedness estimation become increasingly crucial. Traditional approaches for document representation are mostly based on the Vector Space (VSM) Model or the Bag of Words (BOW) model which takes a document as an unordered collection of words and only document-level

This submission is an extended version of the paper published in DaWaK'12, which was selected by the DaWaK'12 program committee for possible publication in the LNCS Transactions on Large-Scale Data- and Knowledge-Centered Systems.

statistical information is recorded (e.g., document frequency, inverse document frequency). Due to the lack of capturing semantics in texts, for certain tasks, especially fine-grained information discovery applications, such as mining relationships between concepts, VSM demonstrates its inherent limitations because of its rationale for computing relatedness between words only based on the statistical information collected from documents themselves. It leads to great semantic loss because terms not appearing in the text literally cannot be taken into consideration.

Our previous work [1] introduced a special case of text mining focusing on detecting semantic relationships between two concepts across documents, which we refer to as Concept Chain Queries (CCQ). A concept chain query involving concepts A and B has the following meaning: find the most plausible relationship between concept A and concept B assuming that one or more instances of both concepts occur in the corpus, but not necessarily in the same document. For example, both may be football lovers, but mentioned in different documents. However, the previous solution was built under the VSM assumption only for the document collection, which limited the scope of the discovered results. For instance, "Albert Gore" is closely related to "George W. Bush" since the two men together produced the most controversial presidential election in 2000, which was the only time in American history that the Supreme Court has determined the outcome of a presidential election. However, "Albert Gore" cannot be identified as a relevant concept to "George W. Bush" if it does not occur in the document collection where the concept chain queries are performed. Furthermore, the semantic relatedness between concepts computed in [1] is solely measured by the statistical information gathered from the corpus such as term frequency (TF), inverse document frequency (IDF), with no background knowledge incorporated.

In this work, we present a new approach that attempts to address the above problems by utilizing background knowledge to provide a better semantic representation of any text and a more appropriate estimation of semantic relatedness between concepts. This is accomplished through leveraging Wikipedia, the world's currently largest human built encyclopedia. Specifically, in addition to inspecting the given documents, we sift through the articles and anchor texts in the space of Wikipedia, attempting to integrate relevant background knowledge for the topics being searched. Our algorithm is motivated by the Explicit Semantic Analysis (ESA) [3] technique where ESA maps a given text or concept to a conceptual vector space spanned by all Wikipedia articles, and thus rich background knowledge can be integrated into the semantic representation of that text or concept. Here we adapt and improve the ESA model in two dimensions. First, we attempt to identify only the most relevant concepts generated from ESA for topic semantic representation and relatedness computation through introducing a series of heuristic steps for noise reduction. Second, we go one step further to take into account anchor texts inside relevant Wikipedia articles. This is based on the observation that the anchor texts within an article are usually highly relevant to it. Therefore, if an article is identified to be relevant to our search topic, it is highly likely that its anchors are topic-relevant as well. To validate the proposed techniques, a significant amount of queries covering different scenarios were conducted. The results showed that through incorporating Wikipedia knowledge, the most relevant concepts to the given topics were ranked in top positions.

Our contribution of this effort can be summarized as follows: (1) a new Wiki-enabled cross-document knowledge discovery framework has been proposed and implemented which effectively complements the existing information contained in the document collection and provides a more comprehensive knowledge representation and mining framework supporting various query scenarios; (2) effective noise filtering techniques are provided through developing a series of heuristic strategies for noise reduction, which further increases the reliability of the overall knowledge encoded; (3) to the best of our knowledge, little work has been done to consider ESA as an effective aid in cross-document knowledge discovery. In this work, built on the traditional BOW representation for corpus content analysis, the ESA technique has been successfully integrated into the discovery process and a better estimation of semantic relatedness is provided by combining various evidences from Wikipedia such as article content and anchor texts. We envision this integration would also benefit other related tasks such as question answering and cross–document summarization; (4) the proposed approach presents a new perspective of alleviating semantic loss caused by only using the Vector Space Model (VSM) on the corpus level through incorporating relevant background knowledge from Wikipedia; (5) in addition to uncovering "what relationships might exist between two topics of interest", our method further explores another dimension of the analysis by generating evidence trails from Wikipedia to interpret the nature of the potential concept relationships.

The remainder of this paper is organized as follows: Sect. 2 describes related work. Section 3 introduces concept chain queries. In Sect. 4, we present our proposed method of utilizing Wikipedia knowledge for answering concept chain queries. Experimental results are presented and analysed in Sect. 5, and is followed by the conclusion and future work.

2   Related Work

Mining semantic relationships/associations between concepts from text is important for inferring new knowledge and detecting new trends. Built within the discovery framework established by Swanson and Smalheiser [4], Srinivasan proposed the open and closed text mining algorithm [2] to automatically discover interesting concepts from MEDLINE. There has also been work on discovering connections between concepts across documents using social network graphs, where nodes represent documents and links represent connections (typically URL links) between documents. However, much of the work on social network analysis has focused on special problems, such as detecting communities [7, 11]. Our previous work [1] introduced Concept Chain Queries (CCQ), a special case of text mining focusing on detecting cross-document links between concepts in document collections, which was motivated by Srinivsan's closed text mining algorithm [4]. Specifically, the solution proposed attempted to generate concept chains based on the "Bag of Words" (BOW) representation on the text corpus and extended the technique in [2] by considering multiple levels of interesting concepts instead of just one level as in the original method. Each document in [1] was represented as a vector containing all the words appearing in the relevant text snippets in the corpus but did not take any auxiliary knowledge into consideration, whereas in this new solution, in addition to corpus level content analysis, we further examine the potential of integrating the Explicit Semantic Analysis (ESA) [3] technique to better serve this task which effectively incorporates more comprehensive knowledge from Wikipedia. There have been a lot of efforts in earlier research as discussed in [31], trying to add semantics to traditional VSM based text processing. Deerwester [32] introduced Latent Semantic Indexing (LSI) for automatic identification of concepts using singular value decomposition. However, it has been found that LSI can rarely improve the strong baseline established by SVM [5, 35, 36]. This becomes part of our motivations of integrating ESA in this work.

WordNet, a lexical database for the English language [18], has been widely used to overcome the limitations of the VSM in text retrieval [19], document clustering [20, 21] and document categorization [22, 23]. For example, Hotho et al. [6] utilized WordNet to improve the VSM text representation and Scott et al. [9] proposed a new representation of text based on WordNet hypernyms. These WordNet-based approaches were shown to alleviate the problems of BOW model but are subject to relatively limited coverage compared to Wikipedia, the world's largest knowledge base to date. Gurevych et al. used Wikipedia to integrate semantic relatedness into the information retrieval process [24], and Müller et al. [25] used Wikipedia in domain-specific information retrieval. Gabrilovich et al. [5] applied machine learning techniques to Wikipedia and proposed a new method to enrich document representation from this huge knowledge repository. Specifically, they built a feature generator to identify most relevant Wikipedia articles for each document, and then used concepts corresponding to these articles to create new features. As claimed in [5], one of the advantages using Wikipedia over Open Directory Project (ODP) is the articles in Wikipedia are much cleaner than typical Web pages, and mostly qualify as standard written English. However, without proper feature selection strategies employed, there will still be a large amount of noise concepts introduced by the feature generator. Another concern needing to be drawn here is the challenge of efficiently processing large scale data. Bonifati and Cuzzocrea [28] presented a novel technique to fragment large XML documents using structural constraints such as size, tree-width, and tree-depth. Cuzzocrea et al. [29] used K-means clustering algorithm to perform the fragmentation of very large XML data warehouses at scale. Cuzzocrea and Bertino [30] proposed a framework for efficiently processing distributed collections of XML documents. While in this work, we import the XML dump of Wikipedia into relational database and build multi-level indices on the database to support efficient queries against Wiki data.

Serving as an integral part of information retrieval and natural language processing, semantic similarity estimation between words has gained increasing attention over the past years. Various web resources have been considered for this purpose [14–16]. Rinaldi [34] proposed a metric to compute the semantic relatedness between words based on a semantic network built from ontological information. Bollegala [17] developed an automatic method for semantic similarity calculation using returned page counts and text snippets generated by a Web search engine. Gabrilovich et al. also [3] presented a novel method, Explicit Semantic Analysis (ESA), for fine-grained semantic representation of unrestricted natural language texts. Using this approach, the meaning of any text can be represented as a weighted vector of Wikipedia-based concepts (articles), called an interpretation vector [3]. Gabrilovich et al. [3] also discussed the problem of possibly containing noise concepts in the vector, especially for text fragments containing multi-word phrases (e.g., multi-word names like George Bush). Our proposed solution is motivated by this work and to tackle the above problems we have developed a sequence of heuristic strategies to filter out irrelevant concepts and clean the vector. Another interesting work is an application of ESA in a cross-lingual information retrieval setting to allow retrieval across languages [8]. In that effort the authors performed article selection to filter out those irrelevant Wikipedia articles (concepts). However, we observed the selection process resulted in the loss of many dimensions in the following mapping process, whereas in our proposed approach, the process of article selection is postponed until two semantic profiles have been merged so that the semantic loss could be possibly reduced to the minimum. Furthermore, in comparison to [13, 27], we also tap into another valuable information resource, i.e. the Wikipedia anchor texts, along with articles to provide better semantic relatedness estimation.

3   Concept Chain Queries

As described earlier, concept chain query (CCQ) is attempting to detect links between two concepts (e.g., two person names) across documents. A concept chain query involving concept A and concept B intends to find the best path linking concept A to concept B. The paths found stand for potential conceptual connections between them. Figure 1 gives an example of CCQ, where the query pair is "Nashiri :: Nairobi attack". Since "Nashiri" co-occurs with "Jihad Mohammad Ali al Makki" in the same sentence in Document 1, and "Nairobi attack" co-occurs with "Jihad Mohammad Ali al Makki" in the same sentence in Document 2, "Nashiri" and "Nairobi attack" can be linked through the concept "Jihad Mohammad Ali al Makki".

Document 1:
Nashiri and *his cousin*, *Jihad Mohammad*, returned to Afghanistan, probably in 1997, Nashiri again encountered Bin Ladin, still recruiting for "the coming battle with the United States." Nashiri joined al Qaeda and later was recognized as the chief of al Qaeda operations in and around the Arabian Peninsula.
Document 2:
In late 1998, al Qaeda decided mounting an attack against a U.S. vessel and *Jihad Mohammad*, also known as *Azzam*, was a suicide bomber for the *Nairobi attack*.

Fig. 1. A concept chain example for the query "Nashiri :: Nairobi attack"

3.1   Semantic Profile for Topic Representation

A semantic profile is essentially a set of concepts that together represent the corresponding topic. To further differentiate between the concepts, semantic type (ontological information) is employed in profile generation. The concept mapping process is basically a two-step task: (1) we extract concepts from the document collection using

Semantex [10]; (2) the extracted concepts are mapped the counterterrorism domain ontology [1]. Table 1 illustrates part of semantic type – concept mappings.

Table 1. Semantic type - concept mapping

Semantic type	Instances
Religion	Islam, Muslim
Human action	attack, killing, covert action, international terrorism
Leader	vice president, chief, governor
Country	Iraq, Afghanistan, Pakistan, Kuwait
Infrastructure	World Trade Centre
Diplomatic building	consulate, pentagon, UAE Embassy

Thus each profile is defined as a vector composed of a number of semantic types.

$$profile(T) = \{ST_1, ST_2, \ldots, ST_n\} \qquad (1)$$

Where ST_i represents a semantic type to which concepts appearing in the topic-related text snippets belong. We used sentence as window size to measure relevance of appearing concepts to the topic term. Under this representation each semantic type is again referred to as an additional level of vector composed of a number of terms that belong to this semantic type.

$$ST_i = \{w_{i,1}m_1, w_{i,2}m_2, \ldots, w_{i,n}m_n\} \qquad (2)$$

Where m_j represents a concept belonging to semantic type ST_i, and $w_{i,j}$ represents its weight under the context of ST_i and sentence level closeness. When generating the profile we replace each semantic type in (1) with (2).

In (2), to compute the weight of each concept, we employ a variation of $TF*IDF$ weighting scheme and then normalize the weights:

$$w_{i,j} = s_{i,j} \, / \, highest(s_{i,l}) \qquad (3)$$

Where $l = 1, 2, \ldots, r$ and there are totally r concepts for ST_i, $s_{i,j} = df_{i,j}*Log(N/df_j)$, where N is the number of sentences in the collection, df_j is the number of sentences concept m_j occurs, and $df_{i,j}$ is the number of sentences in which topic T and concept m_j co-occur and m_j belongs to semantic type ST_i. By using the above three formulae we can build the corresponding profile representing any given topic.

3.2 Concept Chain Generation

We adapt Srinivasan's closed discovery algorithm [2] to build concept chains for any two given topics. Each concept chain generated reveals a plausible path from concept A to concept C (suppose A and C are two given topics of interest). The algorithm of generating concept chains connecting A to C is composed of the following three steps.

1. Conduct independent searches for A and C. Build the A and C profiles. Call these profiles AP and CP respectively.
2. Compute a B profile (BP) composed of terms in common between AP and CP. The weight of a concept in BP is the sum of its weights in AP and CP. This is the first level of intermediate potential concepts.
3. Expand the concept chain using the created BP profile together with the topics to build additional levels of intermediate concept lists which (i) connect the topics to each concept in BP profile in the sentence level within each semantic type, and (ii) also normalize and rank them (as detailed in Sect. 3.1).

4 Wikipedia as an Information Resource

Wikipedia is currently the largest human built encyclopedia in the world. It has over 5,000,000 articles by April 05, 2011, and is maintained by over 100,000 contributors from all over the world. As of February 2013, there are editions of Wikipedia in 285 languages. Knowledge in Wikipedia ranges from psychology, math, physics to social science and humanities. To utilize Wikipedia knowledge to complement the existing information contained in the document collection, two important information resources, Wikipedia article contents and anchor texts are considered. Specifically, appropriate content and link analysis will be performed on Wikipedia data and the mined relevant knowledge will be used to further improve our query model and semantic relatedness estimation module.

4.1 Semantic Relatedness Measures

Semantic relatedness indicates degree to which words are associated via any type (such as synonymy, meronymy, hyponymy, hypernymy, functional, associative and other types) of semantic relationships [37]. The measures of computing semantic relatedness between concepts can be grouped into four classes in general [33]: the path length based measures that use the length of path connecting concepts in the taxonomy to measure the closeness between concepts; the information content based measures that rely on the shared information content between concepts; the feature based measures that exploit the common characteristics of concepts; and the hybrid measures that combine the previous three measures. The similarity measures defined in this work can be viewed as an extension of the information content based measure.

4.2 Article Content Analysis

For content analysis, we have adapted the Explicit Semantic Analysis (ESA) technique proposed by Gabrilovich et al. [3] as our underlying content-based measure for analyzing Wikipedia articles relevant to the given topics of interest. In ESA, each term (e.g., topic of interest) is represented by a concept vector containing relevant concepts (Wikepedia articles) to the topic along with their association strengths and each text fragment can also be mapped to a weighted vector of Wikipedia concepts called an interpretation vector. Therefore, computing semantic relatedness between any two text fragments can be naturally transformed into computing the Cosine similarity between interpretation vectors of two texts.

Using the ESA method, each article in Wikipedia is treated as a Wikipedia concept (the title of an article is used as a representative concept to represent the article content), and each document is represented by an interpretation vector containing related Wikipedia concepts (articles) with regard to this document. Formally, a document d can be represented as follows:

$$\phi(d) \; = \; < as(d, a_1), \ldots, as(d, a_n) > \tag{4}$$

Where $as(d, a_i)$ denotes the association strength between document d and Wikipedia article a_i. Suppose d is spanned by all words appearing in it, i.e., $d = < w_1, w_2, \ldots, w_j >$, and the association strength $as(d, a_i)$ is computed by the following function:

$$as(d, a_i) = \sum_{w_j \in d} tf_d(w_j) tf \bullet idf_{a_i}(w_j) \tag{5}$$

Where $tf_d(w_j)$ is the occurrence frequency of word w_j in document d, and $tf \bullet idf_{a_i}(w_j)$ is the $tf{\cdot}idf$ value of word w_j in Wikipedia article a_i. As a result, the vector for a document is represented by a list of real values indicating the association strength of a given document with respect to Wikipedia articles. By using efficient indexing strategies such as single-pass in memory indexing, the computational cost of building these vectors can be reduced to within 200–300 ms. In concept chain queries, the topic input is always a single concept (a single term or phrase), and thus Eq. (5) can be simplified as below as $tf_d(w_j)$ always equals 1:

$$as(d, a_i) = \sum_{w_j \in d} tf \bullet idf_{a_i}(w_j) \tag{6}$$

As discussed above, the original ESA method is subject to the noise concepts introduced, especially when dealing with multi-word phases. For example, when the input is *Angelina Jolie*, the generated interpretation vector will contain a fair amount of noise concepts such as *Eudocia Angelina*, who was the queen consort of Stephen II Nemanjić of Serbia from 1196 to 1198. This Wikipedia concept (article) is selected and ranked high in the interpretation vector because the term Angelina occurs many times in the article "*Eudocia Angelina*", but obviously this article is irrelevant to the given topic *Angelina Jolie*.

In order to make the interpretation vector more precise and relevant to the topic, we have developed a sequence of heuristics to clean the vector. Basically, we use a modified Levenshtein Distance algorithm to measure the relevance of the given topic to each Wikipedia concept generated in the interpretation vector. Instead of using allowable edit operations of a single character to measure the similarity between two strings as in the original Levenshtein Distance algorithm, we view a single word as a unit for edit operations, and thus the adapted algorithm can be used to compute the similarity between any two text snippets. The heuristic steps used to remove noise concepts are illustrated in Fig. 2.

Input: a topic T of interest
      an interpretation vector V representing the topic T

Output: a cleaned Wikipedia-based concept vector V' representing the topic T

    1. If T is a single word topic, then count the number of occurrences of T in the article texts represented by each concept v_i in V, respectively. If T occurs more than 3 times, then keep v_i in V, otherwise, remove v_i from V.

    2. If T is a multi-word topic, then the adapted Levenshtein distance algorithm applies to measure the relevance of each Wikipedia concept (article) v_i in V to topic T.

      2.1. If NumOfWords(T) $\leq$ 2, then extract all text snippets TS_j within the window size NumOfWords(T)+1 from the article text of v_i. If there exists a j such that LevenshteinDistance(T, TS_j) <= 1, then keep v_i in V, otherwise, remove v_i from V.

      2.2. If NumOfWords(T) > 2, then extract all text snippets TS_j within the window size NumOfWords(T)+2 from the article text of v_i. If there exists j such that LevenshteinDistance(T, TS_j) $\leq$ 2, then keep v_i in V, otherwise, remove v_i from V.

Fig. 2. The interpretation vector cleaning procedure

4.3   Article Link Analysis

Anchor texts, another type of valuable information resource provided by Wikipedia in addition to the textual content of articles, imply rich hidden associations between different Wikipedia concepts. For example, the Wikipedia article talking about "*Osama bin Laden*" contains a great number of potential terrorists who are related to him and terrorism events that he was involved in (appearing as anchor texts in the article). Therefore, through inspecting anchor texts in each relevant Wikipedia article, we are able to find a fair amount of interesting concepts related to the topic. Figure 3 gives part of the anchors in the article "*Osama bin Laden*".

We assume that two concepts (articles) sharing similar anchors may be closer to each other in terms of semantic relatedness. As discussed earlier, given a topic of interest, we can represent it as an interpretation vector containing the relevant Wikipedia articles using the ESA method. Also, each Wikipedia article can be further represented by the anchors appearring in it. Therefore, we can build an additional vector, called anchor vector, based on the interpretation vector produced for a given search topic. Simiarly, we can approach the semantic relatedness between two topics from another perspective by calculating the Cosine score of the two anchor vectors built for them.

Formally, suppose the interpretation vector for a topic T_i is $V_i = < article_1, article_2, ..., article_m >$, where $article_i$ in V_i represents a Wikipedia article relevant to T_i, then the topic T_i can be further represented as an *Anchor Vector (AV)* as follows.

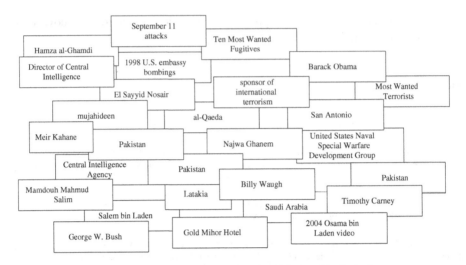

Fig. 3. Wikipedia anchors related to *"Osama bin Laden"*

$$AV(T_i) = << w_{i,1,1}anchor_{1,1}, w_{i,2,1}anchor_{2,1}, \ldots >,$$
$$\ldots, \tag{7}$$
$$< w_{i,1,m}anchor_{1,m}, w_{i,2,m}anchor_{2,m}, \ldots >>$$

Where $anchor_{x,y}$ represents the anchor text $anchor_x$ appearing in $article_y$ in V_i, and $w_{i,x,y}$ is the weight for $anchor_{x,y}$. To calculate $w_{i,x,y}$, we count the number of sub-vectors within $AV(T_i)$ in which $anchor_{x,y}$ appears, and then normalize it:

$$w_{i,x,y} = \frac{w_{i,x,y}}{highest(w_{i,d,y})} \tag{8}$$

Where $d = 1,2,\ldots,r$ and there are totally r anchors in Wikipedia. Therefore, the semantic relatedness between two topics of interest can be estimated as follows:

$$Sim(T_i, T_j) = Cosine(AV(T_i), AV(T_j)) \tag{9}$$

4.4   Integrating Wikipedia Knowledge into Concept Chain Queries

Given the advantages of using Wikipedia as an effective information aid for semantic representation, we integrate the knowledge derived from Wikipedia into our concept chain queries. Specifically, we build interpretation vectors (using our adapted ESA method) and anchor vectors (using the method described in Sect. 4.2) for both the two given topics and each intermediate concept in the merged BP profile, and then compute the Cosine similarities between the topics and each concept in the BP profile using the corresponding interpretation vectors and anchor vectors, respectively. The final ranking will be an integrated scheme considering the following three types of similarities.

Corpus-level TF*IDF-based Similarity. As the most widely used document representation, the BOW representation has demonstrated its advantages. It is simple to compute and strictly sticking to the terms occurring in the document, thereby preventing outside noise concepts that do not appear in the document from flowing into the feature space of the representation. Given these benefits, a variation of *TF*IDF* weighting scheme under the context of BOW representation is incorporated into our final ranking to capture corpus level statistical information. We call this kind of similarity the TF*IDF-based similarity.

ESA-based Similarity. Unlike the BOW model, ESA makes use of the knowledge outside the documents themselves to compute semantic relatedness. It well compensates for the semantic loss resulted from the BOW technique. The relatedness between two concepts in ESA is computed using their corresponding interpretation vectors containing related concepts derived from Wikipedia. In the context of concept chain queries, we compute the Cosine similarity between the interpretation vectors of topic A and each concept in the intermediate BP profile, as well as between topic C and each concept V_i, and take the average of two Cosine similarities as the overall similarity for each concept V_i in BP. We call this kind of similarity the ESA-based similarity.

Anchor-based Similarity. Anchor texts have served as another important information aid in our algorithms to provide highly relevant concepts to the given topics through considering the descriptive or contextual information for relevant Wikipedia articles. As with the case of computing the ESA-based similarity for topic A(C) and each concept V_i in the intermediate BP profile using the interpretation vectors, here anchor vectors are used to measure concept closeness. We refer to this type of similarity the Anchor-based similarity.

Integrating TF*IDF-based Similarity, ESA-based Similarity and Anchor-based Similarity into the Final Ranking. The TF*IDF-based similarity, ESA-based similarity and Anchor-based similarity are finally combined to form a final ranking for concepts generated in the intermediate profiles:

$$S_{overall} = (1 - \lambda_1 - \lambda_2)S_{TFIDF} + \lambda_1 S_{ESA} + \lambda_2 S_{anchor} \tag{10}$$

Where λ_1 and λ_2 are two tuning parameters that can be adjusted based on the preference on the three similarity schemes in the experiments. S_{TFIDF} refers to the TF*IDF-based similarity, S_{ESA} the ESA-based similarity, and S_{anchor} the Anchor-based similarity.

4.5   Annotating Semantic Relationships Between Concepts

In addition to answering "what relationships might exist between two topics?", we go one step further to collect relevant text snippets extracted from multiple Wikipedia articles in which the discovered chains appear. This is in fact a multi-document summary that explains the plausible relationship between topics with intensive knowledge derived from Wikipedia. For example, given a query pair: "*Bin Laden*" and "*Abdel-Rahman*", one of the discovered concept chains is: *Bin Laden → Azzam → Abdel-Rahman*. Our goal now is to find supporting evidence that interprets how

"*Bin Laden*" is linked to "*Abdel-Rahman*" through "*Azzam*" in the space of Wikipedia. We consider this process as the chain-focused sentence retrieval problem and decompose it into the following two subtasks.

Chain-Relevant Article Retrieval. This subtask takes a generated concept chain as input and attempts to find relevant Wikipedia articles for it. One important criterion that needs to be met is the identified Wikepeida articles should be relevant to the whole chain (i.e. relevance to the given topics (end points of the chain) as well as intervening concepts), not just to any individual segment of the chain. To achieve this, we first (1) build the corresponding interpretation vectors for all of the concepts appearing in the chain; (2) perform noise removal using the cleaning procedure described in Fig. 2; (3) construct a ranked list of Wikipedia articles by intersecting the resulting interpretation vectors with each article weighted using formula 6; (4) follow similar steps as above to construct a ranked list of anchors (note that an anchor also represents a Wikipedia article) with each anchor weighted using formula 8. The articles represented by the concepts in the two ranked lists are viewed as chain relevant articles.

Chain-Focused Sentence Retrieval. This step inspects the content of each article generated from the previous step and extracts sentences that explain each segment of chain. For example, the chain *Bin Laden* → *Azzam* → *Abdel-Rahman* is composed of two segments: *Bin Laden* → *Azzam* and *Azzam* → *Abdel-Rahman*. For the segment *Bin Laden* → *Azzam*, sentences where "*Bin Laden*" and "*Azzam*" co-occur will be extracted as supporting evidence for this partial chain. Figure 4 shows the generated evidence trail for this example.

Supporting Evidence for **Bin Laden** and **Azzam**
In 1989, after the Soviets pulled out of Afghanistan, **Azzam** *and his deputy Osama* **bin Laden** *decided to keep their movement permanent and founded the Al Qaeda.*

Supporting Evidence for **Azzam** and **Abdel-Rahman**
During theological studies in Egypt, **Azzam** *met Omar* **Abdel-Rahman***, Dr.Ayman al-Zawahiri and other followers of Sayyed Qutb, an extremely influential leader of the Egyptian Muslim Brotherhood, who had been executed by President Gamal Abdel Nasser in 1966.*

Fig. 4. Evidence trail generated from Wikipedia articles for the concept chain *Bin Laden* → *Azzam* → *Abdel-Rahman*

4.6   The New Mining Model

To summarize, the new model of answering concept chain queries consists of two sequential steps as shown in Figs. 5 and 6. Figure 5 illustrates the first step which discovers potential relationships between two given topics from the given document collection without background knowledge incorporated. Figure 6 details how Wikipedia knowledge is integrated into this discovery process and facilitates better estimation of semantic relatedness between concepts. Also, we go one step further and require the response to be a set of Wikipedia text snippets (i.e. evidence trail) in which

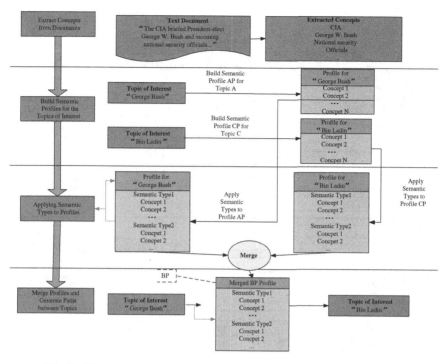

Fig. 5. The new model of answering concept chain queries: component-1

the discovered concept chain occurs. This may assist a user with the second dimension of the analysis process, i.e. when the user has to peruse the documents to figure out the nature of the relationship underlying a suggested chain.

5   Empirical Evaluation

A challenging task for the evaluation was constructing an evaluation data set, since there are no standard data sets available for quantitatively evaluating concept chains. We performed our evaluation using the 9/11 counterterrorism corpus. The Wikipedia snapshot used in the experiments was dumped on April 05, 2011.

5.1   Processing Wikipedia Dumps

As an open source project, the entire content of Wikipedia is easily obtainable. All the information from Wikipedia is available in the form of database dumps that are released periodically, from several days to several weeks apart. The version used in this work was released on April 05, 2011, which was separated into 15 compressed XML files and totally occupies 29.5 GB after decompression, containing articles, templates, image descriptions, and primary meta-pages. We leveraged MWDumper [12] to import the XML dumps into our MediaWiki database, and after the parsing process, we identified 5,553,542 articles.

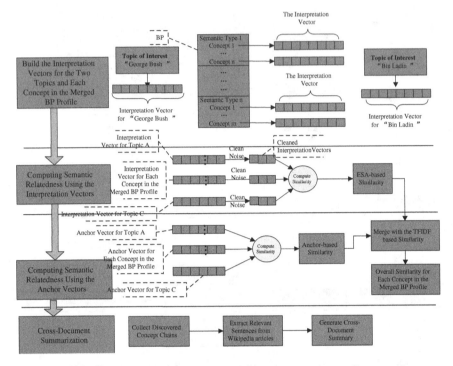

Fig. 6. The new model of answering concept chain queries: component-2

5.2   Evaluation Data

We performed concept chain queries on the 9/11 counterterrorism corpus. This involves processing a large open source document collection pertaining to the 9/11 attack, including the publicly available 9/11 commission report. The report consists of Executive Summary, Preface, 13 chapters, Appendix and Notes. Each of them was considered as a separate document resulting in 337 documents. The whole collection was processed using Semantex [10] and concepts were extracted and selected as shown in Table 1. Query pairs covering various scenarios (e.g., ranging from popular entities to rare entities) were selected by the assessors and used as our evaluation data. We selected chains of lengths ranging from 1 to 4 in terms of the number of associations. The chains were selected by going through the same procedure as in [26], which is also described as follows:

1. We ran queries with various pairs of topics: in the counterterrorism corpus, the topics were mostly named entities.
2. For each topic pair, the relevant paragraphs for either topic were then manually inspected: we selected those where there was a logical connection between the two topics.
3. After achieving agreement among all annotators, we then generated the concept chains for these topic pairs (and paragraphs) as evaluation data.

The above process generated 37 chains in 9/11 corpus which will be used as truth chains for later experiments.

5.3 Experimental Results

Parameter Settings. As mentioned in Sect. 4.3, a combination of TF*IDF-based similarity, ESA-based similarity and Anchor-based similarity is used to rank the links detected by our system. λ_1 and λ_2 in Eq. 10 are two parameters that need to be tuned so that the generated similarity between two concepts best matches the judgements from our assessors. To accomplish this, we first built a set of training data composed of 10 query pairs randomly selected from the evaluation set, and then generated BP profiles for each of them using our proposed method. Among each BP profile, we selected the top 5 concepts (links) within each semantic type, and compared their rankings with the assessors' judgements. The values of λ_1 and λ_2 were tuned in the range of [0.1, 1]. Specifically, we set $\lambda_2 = 0$ or $\lambda_1 = 0$ to evaluate the contribution of each individual part (the ESA-based similarity or the Anchor-based similarity) in the final weighting scheme. When $\lambda_1 \neq 0$ and $\lambda_2 \neq 0$, the best performance was obtained when $\lambda_1 = 0.4$ and $\lambda_2 = 0.3$. These settings were also used in our later experiments.

Query Results. Before proceeding to the evaluation of the proposed model, we first conducted an experiment to demonstrate the improved performance of our adapted ESA method against the original ESA. We selected 10 concepts that we have good knowledge about as shown in Table 2 and then built the interpretation vectors for each of them using the original ESA and our adapted ESA respectively. We calculated the averaged precision defined as below to measure the performance of the two approaches.

$$aveP = \left(\sum_{i=1}^{N} \frac{concept\ found\ and\ relevant}{total\ concepts\ found} \right) / N \qquad (11)$$

Table 2. Ten concepts used for the interpretation vector construction

Semantic type	Belonging concept
Person	George Bush
	Bill Clinton
Organization	Central Intelligence Agency
	United States Federal Government
Event	World War
	September 11 attacks
	Lewinsky Scandal
Science	Data Mining
	Natural Language Processing
	Artificial Intelligence

where N is the number of concepts under consideration. The results are illustrated in Fig. 7 where the X-axis indicates the number of concepts kept in each of the interpretation vectors and the Y-axis indicates the averaged precision ratio. It is obvious that our adapted ESA achieves significant improvement over the original one for identifying topic-related Wikipedia concepts.

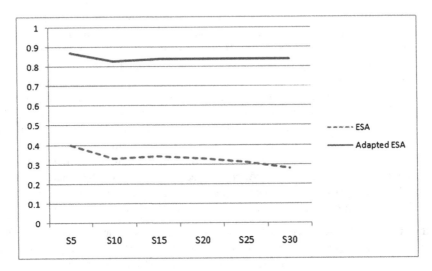

Fig. 7. The Averaged Precision of the generated interpretation vectors using the original ESA and adapted ESA based on processing data in Table 2

Table 3 shows the top 15 concepts generated in the interpretation vectors for 4 sample concepts. For example, for "Lewinsky Scandal", the top 15 concepts in the interpretation vector built using our adapted ESA include most of the people involved in this event in addition to Clinton and Lewinsky themselves, such as Linda Tripp who secretly recorded Lewinsky's confidential phone calls about her relationship with Clinton, and Betty Currie who was the personal secretary of Clinton and well known in the scandal for handling gifts given to Lewinsky by Clinton. However, most of the top concepts identified using the original ESA are representing some irrelevant events.

To further evaluate the performance of the original ESA and the adapted ESA in semantic profile generation, we selected 10 query pairs as shown in Table 4 and generated semantic profiles serving as linking concepts (i.e. BP profile) through selecting common concepts appearing in the two interpretation vectors built for the two given topics. Each concept in the semantic profile was weighted using the original ESA and our adapted ESA respectively. We again calculated the averaged precision to measure the percentage of the relevant concepts in the generated profile. The results are shown in Fig. 8 where the X-axis indicates the number of concepts kept in each generated semantic profile and the Y-axis indicates the averaged precision. It is demonstrated that for BP level semantic profile generation, our adapted ESA also performs much better than the original ESA.

Table 3. Top 15 concepts in the sample interpretation vectors using the adapted ESA and the original ESA

Input	#	Original ESA	Adapted ESA
Data Mining	1	Open-cast_mining	Relational_classification
	2	Opencast_Mining	Relational_data_mining
	3	Mining_engineer	Data_Mining_Extensions
	4	Open_cast_mining	Biological_data
	5	data	Java_Data_Mining
	6	Mine_(industry)	Weather_Data_Mining
	7	Open-cast_mine	National_Center_for_Data_Mining
	8	Golden_Source_of_data	Privacy_preserving_data_mining
	9	Data_withholding	Structure_mining
	10	Data_Havens	Oracle_Data_Mining
	11	Data_Warehousing	Cross_Industry_Standard_Process_for_Data_Mining
	12	Data_Transfer	Knowledge_discovery
	13	Data_rate_(disambiguation)	Data_Pre-processing
	14	Data_General_One	Data_mining_agent
	15	Data_matrix_(disambiguation)	Sequence_mining
Central Intelligence Agency	1	Agency_(disambiguation)	United_States._Central_Intelligence_ Agency
	2	United_States._Central_Intelligence_Agency	Central_Intelligence_Agency_Museum
	3	Starfleet_Intelligence	Central_Intelligence_Agency_library
	4	Nigerian_intelligence	The_Agency
	5	Virginia_farmboys	National_Intelligence_Agency_(United_States)
	6	Directorate_for_Inter-Service_Intelligence	Agency
	7	Process_of_intelligence	Office_of_Scientific_Intelligence
	8	14th_Intelligence_Company	Intelligence_officer
	9	Intelligence_augmentation	Security_agency
	10	Human_intelligence_ (disambiguation)	John_N._McMahon
	11	Israeli_Intelligence_Agency	National_Intelligence_Board
	12	Agência_Brasileira_de_ Inteligência	Director_of_the_Central_Intelligence_Agency
	13	Central_(disambiguation)	Military_Intelligence_Division
	14	Administrative_agency	Private_intelligence_agency
	15	Job_agency	Intelligence_agency
Lewinsky Scandal	1	Scandal-mongering	Clinton:_His_Struggle_with_Dirt
	2	HIV-tainted-blood_scandal	Monica_Lewinsky
	3	Scandal_of_Scientology	Lewinsky_scandal
	4	The_Scandal_of_Scientology_ (book)	Linda_Tripp
	5	Iraq_War_Scandal_ (disambiguation)	Susan_Schmidt
	6	CDU_contribution_scandal	Kramerbooks_&_Afterwords
	7	Parmalat_scandal	Betty_Currie
	8	Coingate	Monica
	9	Black_Mist_Scandal	Affair
	10	Scandal_(disambiguation)	Breuer
	11	2006_Reuters_fake_photos_ scandal	Charles_Ruff
	12	Boesky_scandal	Robert_S._Bennett
	13	Panama_scandal	Mark_Whitaker
	14	Sex_scandals	David_Horsey
	15	Shell_Scandal_of_1915	1983_congressional_page_sex_scandal

Table 5 below shows the top 10 concepts generated in the semantic profiles for our sample query pairs: "George Bush :: Al Gore" and "Michael Jordan :: Charles Barkley". For the query pair "Michael Jordan :: Charles Barkley", the top 10 concepts identified as the interlinking terms using our adapted ESA include their most relevant persons, events, etc., and noise concepts are successfully removed from the semantic

Table 4. Ten query pairs used for the semantic profile generation comparison

Topic A	Topic C
George Bush	Al Gore
Michael Jordan	Charles Barkley
Sadam Hussein	Gulf War
Northern Alliance	European Union
Wall Street	New York Times
Steve Jobs	Mark Zuckerberg
Knowledge Discovery	Document Classification
Abdel Rahman	Blind Sheikh
Saudi Arabia	Kuwait
Terrorist Attack	Bill Clinton

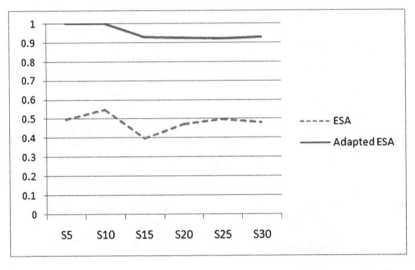

Fig. 8. The Averaged Precision of the Intermediate Semantic Profile (BP profile) Generation using the original ESA and adapted ESA based on processing data in Table 4.

profile and key interlinking concepts are boosted to higher positions such as "I_May_Be_Wrong_but_I_Doubt_It", a memoir by Charles Barkley that recounts some of Barkley's memorable experiences including his involvement with Michael Jordan as a member of the "Dream Team", and "1993_NBA_Finals", the championship round of a historic season when Michael Jordan led the Chicago Bulls to play against the Phoenix Suns which was led by Charles Barkley. By contrast, the original ESA failed to rank high for some very important concepts related to them.

In terms of concept chain queries, we have also conducted a qualitative evaluation of the proposed model for generating various lengths of chains using the precision ratio defined below.

Table 5. Top 10 concepts in the sample semantic profiles generated by the adapted ESA and the original ESA

Input	#	Original ESA	Adapted ESA
George Bush :: Al Gore	1	George_Rose_(disambiguation)	Electoral_history_of_George_W._ Bush
	2	Electoral_history_of_George_ W. _Bush	Al_Gore_presidential_campaign,_ 2000
	3	Sir_Ralph_Gore,_4th_Baronet	Snippy
	4	St_George_Gore-St_George	Non-rigid_designator
	5	Sir_Arthur_Gore,_1st_Baronet	United_States_presidential_election_ in_Massachusetts,_2000
	6	Electoral_history_of_George_H. _W._Bush	High_Performance_Computing_and_Communication_ Act_of_1991
	7	Al_Gore_presidential_campaign, _2000	Millie_(dog)
	8	Tennis_at_the_1908_Summer_ Olympics_–_ _Men's_indoor_doubles	United_States_presidential_election_in_the_District_ of_Columbia,_ 2000
	9	The_Betrayal_of_America	John_Prescott_Ellis
	10	James_Howard_Gore	George_H._W._Bush
Michael Jordan :: Charles Barkley	1	David_Jordan	I_May_Be_Wrong_but_I_Doubt_It
	2	Charles_Blount	1993_NBA_Finals
	3	Charles_Evans	Barkley,_Shut_Up_and_Jam:_Gaiden
	4	Charles_Bronson_(disambiguation)	Best_NBA_Player_ESPY_Award
	5	Charles_Jordan_(magician)	1996_NBA_All-Star_Game
	6	Baggir	1986–87_NBA_season
	7	Manduca_fosteri	1992–93_NBA_season
	8	I_May_Be_Wrong_but_I_Doubt_It	1990–91_NBA_season
	9	Manduca_diffissa	1991–92_NBA_season
	10	Karl_Jordan	Gaiden

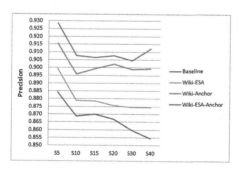

Fig. 9. Search results of chains of length 1

$$precision = \frac{concept\ chains\ found\ and\ correct}{total\ concept\ chains\ found} \qquad (12)$$

Figures 9, 10, 11 and 12 make a comparison of the search results in various models. We have implemented a competitive baseline algorithm (i.e. Srinivasan's 'closed' discovery algorithm) where only the corpus-level TFIDF-based statistical information is considered. In the four figures, the X-axis indicates the number of concepts kept in each semantic type in the search results (S_N means the top N are kept) and the Y-axis indicates the precision values. It is easy to observe that the search performance has been significantly improved with the integration of Wikipedia knowledge, and the best performance is observed when both the Wiki article content and anchor texts are involved.

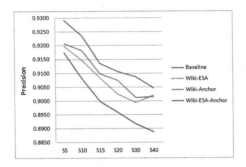

Fig. 10. Search results of chains of length 2

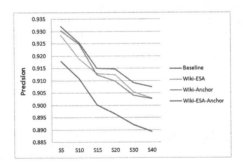

Fig. 11. Search results of chains of length 3

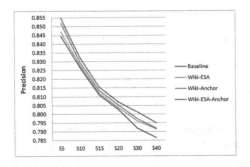

Fig. 12. Search results of chains of length 4

We further used the 37 truth chains described above to measure the performance of the baseline model and various Wiki-enabled models in detecting these chains. In Fig. 13, the X-axis has the same meaning as in Figs. 9, 10, 11 and 12 and the Y-axis now denotes the percentage of the 37 truth chains found by different models. The results also agree with our expectation that the largest percentage of the truth chains were retrieved when incorporating both article content and anchor texts from Wikipedia into the query process.

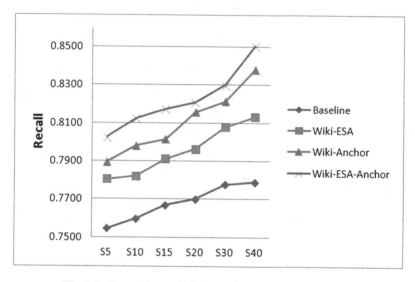

Fig. 13. Comparison of search results using 37 truth chains

Table 6 shows the evidence trails generated for concept chains discovered from Wikipedia. Note that the generated evidence trail is not necessarily from the same Wikipedia article, but could be found through the discovery of knowledge holding across articles. For example, for the concept chain *Betty Ong* → *September 11* → *Mohamed Atta*, sentences were extracted as the supporting evidence from two different Wikipedia articles "*Flight attendant*" and "*American Airlines Flight 11*". Our proposed model successfully found "*Flight attendant*" and "*American Airlines Flight 11*" as two highly relevant Wikipedia articles with regard to "*Betty Ong*" who was a "*Flight attendant*" onboard "*American Airlines Flight 11*" when it was hijacked and flown into the North Tower of the World Trade Center and "*Mohamed Atta*" who was one of the ringleaders of the "*September 11*" attacks, and crashed the "*American Airlines Flight 11*" into the World Trade Center as part of the 9/11 attacks.

Table 6. Evidence trails generated from Wikipedia

Concept chain	Evidence
Betty Ong → September 11 → Mohamed Atta	<u>Sentence 1</u>. The role of flight attendants received heightened prominence after the **September 11** attacks when flight attendants (such as Sandra W. Bradshaw and CeeCee Lyles of United Airlines Flight 93, Robert Fangman of United Airlines Flight 175, Renee May of American Airlines Flight 77 and **Betty Ong** and Madeline Amy Sweeney of American Airlines Flight 11) actively attempted to protect passengers from assault, and also provided vital information to air traffic controllers on the hijackings
	<u>Sentence 2</u>. **Mohamed Atta**, the ringleader of the attacks, and a fellow hijacker, Abdulaziz al-Omari, arrived at Portland International Jetport at 05:41 Eastern Daylight Time on **September 11**, 2001
Rahman → Bin Laden → Al Qaeda	<u>Sentence 1</u>. *Rahman* built a strong rapport with *bin Laden* during the Soviet war in Afghanistan and following Azzam's murder in 1989 Rahman assumed control of the international jihadists arm of MAK/*Al Qaeda*
Gore → Bush → Stephen Hadley	<u>Sentence 1</u>. *Bush*, at the advice of *Hadley*, also proposed greater nuclear arms reductions than *Gore*
Atta → Huffman → Dekkers	<u>Sentence 1</u>. *Atta*, along with Marwan al-Shehhi arrived in Venice, Florida, and visited *Huffman* Aviation to "check out the facility"
	<u>Sentence 2</u>. On the eve of the trial, *Dekkers* sold all of *Huffman*'s holdings minus 10 planes to Triple Diamond, to gather the money needed to repay his business partner

6 Conclusion and Future Work

This paper proposes a new solution for improving cross-document knowledge discovery through our introduced concept chain queries, which focus on detecting semantic relationships between concepts across documents. In this effort, we attempt to incorporate relevant Wikipedia knowledge into the search process, which effectively complements the existing knowledge in document collections and further improves search quality and coverage. Additionally, a better measure for estimating semantic relatedness between terms is devised through integrating various evidence resources from Wikipedia. Experimental results demonstrate the effectiveness of our proposed new approach and show its advantage of alleviating semantic loss caused by only using the Vector Space Model (VSM) on the corpus level.

Future directions include the exploration of other potential resources provided by Wikipedia to further improve query processing, such as infobox information, categories that relevant Wiki articles belong to and the underlying category hierarchy. These valuable information resources may be combined with our defined semantic types to further contribute to ontology modeling. As a cross language knowledge base, we also plan to explore the utilization of Wikipedia knowledge in a cross-lingual setting to better serve different query purposes.

References

1. Jin, W., Srihari, R.K.: Knowledge discovery across documents through concept chain queries. In: Sixth IEEE International Conference on Data Mining Workshops, ICDM Workshops 2006, pp. 448–452. IEEE, December 2006
2. Srinivasan, P.: Text mining: generating hypotheses from MEDLINE. J. Am. Soc. Inform. Sci. Technol. **55**(5), 396–413 (2004)
3. Gabrilovich, E., Markovitch, S.: Computing semantic relatedness using wikipedia-based explicit semantic analysis. In: IJCAI, vol. 7, pp. 1606–1611, January 2007
4. Swanson, D.R., Smalheiser, N.R.: Implicit text linkages between Medline records: using Arrowsmith as an aid to scientific discovery. Libr. Trends **48**(1), 48–59 (1999)
5. Gabrilovich, E., Markovitch, S.: Overcoming the brittleness bottleneck using Wikipedia: enhancing text categorization with encyclopedic knowledge. In: AAAI, vol. 6, pp. 1301–1306, July 2006
6. Hotho, A., Staab, S., Stumme, G.: Wordnet improves text document clustering. In: Proceedings of the Semantic Web Workshop at SIGIR 2003, November 2003
7. Gibson, D., Kleinberg, J., Raghavan, P.: Inferring web communities from link topology. In: Proceedings of the Ninth ACM Conference on Hypertext and Hypermedia: Links, Objects, Time and Space—Structure in Hypermedia Systems: Links, Objects, Time and Space—Structure in Hypermedia Systems, pp. 225–234. ACM, May 1998
8. Sorg, P., Cimiano, P.: Cross-lingual information retrieval with explicit semantic analysis. In: CLEF Workshop 2008 (2008)
9. Scott, S., Matwin, S.: Text classification using WordNet hypernyms. In: Use of WordNet in Natural Language Processing Systems: Proceedings of the Conference, pp. 38–44, August 1998
10. Srihari, R.K., Li, W., Niu, C., Cornell, T.: Infoxtract: a customizable intermediate level information extraction engine. In: Proceedings of the HLT-NAACL 2003 Workshop on Software Engineering and Architecture of Language Technology Systems, vol. 8, pp. 51–58. Association for Computational Linguistics, May 2003
11. Faloutsos, C., McCurley, K.S., Tomkins, A.: Fast discovery of connection subgraphs. In: Proceedings of the Tenth ACM SIGKDD International Conference on Knowledge Discovery and Data Mining, pp. 118–127. ACM, August 2004
12. MWDumper. Software available at http://www.mediawiki.org/wiki/Manual:MWDumper
13. Yan, P., Jin, W.: Improving cross-document knowledge discovery using explicit semantic analysis. In: Cuzzocrea, A., Dayal, U. (eds.) DaWaK 2012. LNCS, vol. 7448, pp. 378–389. Springer, Heidelberg (2012)
14. Jin, W., Srihari, R., Singh, A.: Generating hypotheses from the web. In: Proceedings of the 17th International Conference on World Wide Web, pp. 1211–1212. ACM, April 2008
15. Luo, G., Tang, C., Tian, Y.L.: Answering relationship queries on the web. In: Proceedings of the 16th International Conference on World Wide Web, pp. 561–570. ACM, May 2007
16. Radev, D.R., Libner, K., Fan, W.: Getting answers to natural language questions on the Web. J. Am. Soc. Inform. Sci. Technol. **53**(5), 359–364 (2002)
17. Bollegala, D., Matsuo, Y., Ishizuka, M.: Measuring semantic similarity between words using web search engines. In: WWW 2007, pp. 757–766 (2007)
18. Miller, G.A.: WordNet: a lexical database for English. Commun. ACM **38**(11), 39–41 (1995)

19. Gonzalo, J., Verdejo, F., Chugur, I., Cigarran, J.: Indexing with WordNet synsets can improve text retrieval. arXiv preprint cmp-lg/9808002 (1998)
20. Dave, K., Lawrence, S., Pennock, D.M.: Mining the peanut gallery: opinion extraction and semantic classification of product reviews. In: Proceedings of the 12th International Conference on World Wide Web, pp. 519–528. ACM, May 2003
21. Jing, L., Zhou, L., Ng, M.K., Huang, J.Z.: Ontology-based distance measure for text clustering. In: Proceedings of the Text Mining Workshop, SIAM International Conference on Data Mining (2006)
22. Budanitsky, A., Hirst, G.: Evaluating wordnet-based measures of lexical semantic relatedness. Comput. Linguist. **32**(1), 13–47 (2006)
23. Rodríguez, M.D.B., Hidalgo, J.M.G., Agudo, B.D.: Using WordNet to complement training information in text categorization. arXiv preprint cmp-lg/9709007 (1997)
24. Gurevych, I., Müller, C., Zesch, T.: What to be?-electronic career guidance based on semantic relatedness. In: Annual Meeting-Association for Computational Linguistics, vol. 45(1), p. 1032, June 2007
25. Müller, C., Gurevych, I.: Using wikipedia and wiktionary in domain-specific information retrieval. In: Peters, C., et al. (eds.) CLEF 2008. LNCS, vol. 5706, pp. 219–226. Springer, Heidelberg (2009)
26. Jin, W., Srihari, R.K., Ho, H.H., Wu, X.: Improving knowledge discovery in document collections through combining text retrieval and link analysis techniques. In: Seventh IEEE International Conference on Data Mining, ICDM 2007, pp. 193–202. IEEE, October 2007
27. Yan, P., Jin, W.: Mining semantic relationships between concepts across documents incorporating wikipedia knowledge. In: Perner, P. (ed.) ICDM 2013. LNCS, vol. 7987, pp. 70–84. Springer, Heidelberg (2013)
28. Bonifati, A., Cuzzocrea, A.: Efficient fragmentation of large XML documents. In: Wagner, R., Revell, N., Pernul, G. (eds.) DEXA 2007. LNCS, vol. 4653, pp. 539–550. Springer, Heidelberg (2007)
29. Cuzzocrea, A., Darmont, J., Mahboubi, H.: Fragmenting very large xml data warehouses via k-means clustering algorithm. Int. J. Bus. Intell. Data Min. **4**(3), 301–328 (2009)
30. Cuzzocrea, A., Bertino, E.: A secure multiparty computation privacy preserving OLAP framework over distributed XML data. In: Proceedings of the 2010 ACM Symposium on Applied Computing, pp. 1666–1673. ACM (2010)
31. Turney, P.D., Pantel, P.: From frequency to meaning: vector space models of semantics. J. Artif. Intell. Res. **37**(1), 141–188 (2010)
32. Deerwester, S.: Improving information retrieval with latent semantic indexing. In: Proceedings of the 51st Annual Meeting of the American Society for Information Science, pp. 36–40 (1988)
33. Meng, L., Huang, R., Gu, J.: A review of semantic similarity measures in wordnet. Int. J. Hybrid Inform. Technol. **6**(1), 1–12 (2013)
34. Rinaldi, A.M.: An ontology-driven approach for semantic information retrieval on the web. ACM Trans. Internet Technol. (TOIT) **9**(3), 10 (2009)
35. Wu, H., Gunopulos, D.: Evaluating the utility of statistical phrases and latent semantic indexing for text classification. In: Proceedings of the 2002 IEEE International Conference on Data Mining, ICDM 2003, pp. 713–716. IEEE (2002)
36. Liu, T., Chen, Z., Zhang, B., Ma, W.Y., Wu, G.: Improving text classification using local latent semantic indexing. In: Fourth IEEE International Conference on Data Mining, ICDM 2004, pp. 162–169. IEEE, November 2004
37. Salahli, M.A.: An approach for measuring semantic relatedness between words via related terms. Math. Comput. Appl. **14**(1), 55 (2009)

Author Index

Abdelbaki, Wiem   73

Chen, Qiming   94
Cuzzocrea, Alfredo   115

Dobbie, Gillian   140

Hassan, Ali   20
Hsu, Meichun   94
Huang, David Tse Jung   140

Jiang, Fan   115
Jin, Wei   161

Kim, Jinho   1
Koh, Yun Sing   140

Lee, Suan   1
Lee, Wookey   1
Leung, Carson K.   115
Liu, Dacheng   115

Messaoud, Riadh Ben   73
Moon, Yang-Sae   1

Peddle, Aaron   115
Pedersen, Torben Bach   48

Ravat, Frank   20

Šikšnys, Laurynas   48

Tanbeer, Syed K.   115
Teste, Olivier   20
Thomsen, Christian   48
Tournier, Ronan   20

Yahia, Sadok Ben   73
Yan, Peng   161

Zurfluh, Gilles   20

Printed in the United States
By Bookmasters